WHY DOES THE WORLD STAY GREEN?

Nutrition and survival of plant-eaters

TCR White

CSIRO
PUBLISHING

© TCR White 2005

All rights reserved. Except under the conditions described in the Australian Copyright Act 1968 and subsequent amendments, no part of this publication may be reproduced, stored in a retrieval system or transmitted in any form or by any means, electronic, mechanical, photocopying, recording, duplicating or otherwise, without the prior permission of the copyright owner. Contact **CSIRO** PUBLISHING for all permission requests.

National Library of Australia Cataloguing-in-Publication entry
White, T. C. R. (Thomas C. R.).
   Why does the world stay green? : nutrition and survival of plant-eaters.

   Bibliography.
   Includes index.
   ISBN 0 643 09158 0.

   1. Population biology. 2. Animal-plant relationships. 3. Nitrogen in animal nutrition. I. Title.

   591.53

Available from
CSIRO PUBLISHING
150 Oxford Street (PO Box 1139)
Collingwood VIC 3066
Australia

Telephone:	+61 3 9662 7666
Local call:	1300 788 000 (Australia only)
Fax:	+61 3 9662 7555
Email:	publishing.sales@csiro.au
Web site:	www.publish.csiro.au

Cover photo by istockphoto

Set in 10.5/13 Minion
Cover and text design by James Kelly
Typeset by J & M Typesetting
Printed in Australia by Ligare

The views expressed in this work are the author's own and do not necessarily reflect those of the publisher.

Acknowledgements

This book arose from a series of talks written for Robyn William's 'Occam's Razor' program on ABC Radio National. Only one was broadcast, but my wife said that they provided the nucleus for a natural history book, and kept encouraging me to write it. She also helped with discussion and editing of early drafts, and with proofreading. Many colleagues lent me photographs to illustrate the book; unfortunately, not all of these could be used but I thank them all for their generosity. Finally, my thanks to Ted Hamilton, for his enthusiastic support, and to Anne Findlay, for gentle but beneficial editing.

So here it is, for you, Jan, my Lovely Lady.

Contents

Foreword   *vii*

Chapter 1 The green world   **1**
    Finding food is too hard   4
    Food tastes disgusting or is poisonous   5
    Food is not nutritious enough   6
    But what about the predators?   8
    Nitrogen – the key limiting factor   9
    How herbivores access nitrogen   12

Chapter 2 Herbivores are fussy eaters   **15**
    Seeking out the best: flush-feeders   15
    Going with the flow: seed-eaters   17
    Prolonging the supply: grazers and gall-makers   20
    Creaming off the best: fast-track feeders   23
    Catching the late run: senescence-feeders   26
    Double-dipping   29

Chapter 3 With a little help from microbes   **33**
    Dung-eaters   33
    Detritus-feeders   43

Chapter 4 Meat-eating vegetarians and cannibals   **47**
    Strictly vegetarian?   48
    Starting out carnivorous   51

Opportunistic predators	54
Cannibalism	56

Chapter 5 Feeding the favoured few — 63

Territorial behaviour	63
Social dominance hierarchies	71

Chapter 6 Inefficient killers — 79

Lions and other inefficient killers	80
Bungling invertebrates	84
Food supply is the key	88

Chapter 7 Plagues, outbreaks and the tyranny of weather — 93

Weather's dramatic effects	93
Successful reproductive strategies	100
Weather can affect food quality	104

Afterword	*111*
Further reading	*115*
Index	*119*

Foreword

All biologists worth their salt know that each and every form of life has the capacity to multiply and increase at a truly astonishing, indeed a frightening rate. It is easy to do calculations demonstrating the truth of this. For example, assuming (in all cases) that all descendants survive, one bacterium dividing every 20 minutes would produce approximately 300 grams of bacteria in 24 hours; 150 million tonnes in a month. A female housefly, laying a minimum of 600 eggs in her lifetime, would, at the end of a summer of some eight to 10 generations, have 1.9×10^{20} descendants – or roughly 200 million cubic metres of fly. A female vole reaches sexual maturity in 28 days, has a gestation period of 21 days and produces six to eight young in each litter. In a year she would have a million descendants. By way of contrast, female elephants do not mature sexually until they are 30 years old, have a gestation period of 21 months, and produce an average of only six young in their lifetime. Yet in 750 years one female would have 19 million descendants.

Clearly none of these things happens or the world would be swamped by any one of these creatures. However, sometimes such rates of increase are achieved for brief periods. Then the explosive growth of numbers in a very short time is truly spectacular. Think of plagues of locusts.

When I was an undergraduate I was taught that animals did not increase like this because every species has natural enemies which quickly kill most of the 'surplus' individuals. This is something that seems intuitively obvious – we can easily observe this predation happening all around us in nature. So, on those rare occasions when some animal does reach plague proportions, the assumption is that this must be because something has prevented its natural enemies from regulating its numbers. From this it follows that the way to control populations of pests introduced from another country – be they plant or animal – is to import their natural enemies which had not come with them. When I was a young Forest Entomologist my job centred around these beliefs. It was not until I was confronted with an outbreak of native New Zealand caterpillars defoliating introduced North American pine trees, that I started to doubt this received wisdom. Here was an animal, usually in such low numbers that it is hard to find, either on its natural or adopted hosts, and with a full suite of natural enemies attacking it, suddenly becoming so abundant on these introduced plants that it was destroying them. But it was not doing so on all of them, even when they were in quite close proximity to each other: nor on any of its native food plants. There must be some other explanation. And so there is. However, it took me many years of study and research before I understood what it is.

Perhaps, not surprisingly, it is a very simple explanation. But it is one that is not at all apparent, even to the quite careful observer. The real reason animals do not increase and swamp their environment is because they cannot obtain enough of the sort of food they must have to reproduce and grow. Without this, females can produce few young, and most that are born quickly starve. It is only when, briefly, and for a variety of reasons, there is an increase in the availability of such food that more animals survive. Then, if this increase of their food is large and sustained, we observe plagues and outbreaks.

This book explains how all this comes about in nature and describes some of the many 'ingenious' ways in which animals have evolved to cope with this usually chronic shortage of an essential resource.

If you are like many people – and especially if you watch 'tooth and claw' natural history documentaries on television – you will doubt me. So, too, do many professional biologists. But not, interestingly, those scientists whose work is connected with the nutrition and growth of laboratory and farm animals and birds. Nor do farmers who raise animals for a living. Frequently the response of such people is 'So, what's new?' But I hope that if you are a doubter, you will read what I have to say rather than dismiss it without considering the evidence. Then, perhaps, you may be sufficiently motivated to start looking more closely for yourself at what *really* goes on in this wonderful, if harsh and pitiless world of ours.

Above and beyond all this, however, I think you will be – as I have always been – fascinated and captivated by the many marvellous ways in which animals have evolved to survive in this inadequate world.

T.C.R. White
School of Agriculture and Wine
Waite Agricultural Research Institute
The University of Adelaide
March 2005

1

The green world

Fifty years ago, as a young Forest Entomologist, I visited some of the great balsam fir forests of Canada when they were being attacked by spruce budworm caterpillars. Whole forests were being totally stripped of foliage and nearly all the trees over huge areas were being killed. Only a massive program of aerial spraying with insecticide prevented the death of many more. Some years later I witnessed the same thing happening to plantations of mature pine trees in New Zealand. This time native caterpillars had suddenly found these introduced trees to their taste. There were so many caterpillars eating the needles that, when standing inside the plantation, I was constantly showered with a fine rain of their droppings and could hear them pattering on the forest floor. Some areas needed spraying to save the trees, but most, without being sprayed, subsequently put on new growth that was not attacked. The plantations were again green and healthy, and caterpillars were again few and hard to find.

These two incidents are far from isolated examples. Somewhere in the world there will always be similar attacks taking place. Yet for most of the time, in most places, forests stay green and healthy. 'Why is this so?'

From time to time in many parts of the world, great plagues of locusts will descend, apparently from nowhere, and strip every last vestige of green from the landscape. Mostly, however, locusts are rare and hard to find, like the forest defoliators.

Looking yet further afield, we see that wherever there are plants growing there are all kinds of animals eating them. Everything from large mammals to tiny insects can be seen at all times, and everywhere, spending most of their lives eating plants. And there really are vast numbers of these animals. The huge herds of mammals grazing on African grasslands are a good example. Less obvious, but even more plentiful, are the armies of insects constantly eating every sort of plant. Again, every so often one or other of these herbivores will destroy most or all of their food plants; but for most of the time they do not. On average, herbivores consume only some 7 to 18 per cent of all

Figure 1.1 Canadian balsam fir trees defoliated by spruce budworm caterpillars covered with the silken thread left by the caterpillars lowering themselves to the ground to pupate. Photo courtesy of Canadian Forest Service.

the world's plant production. So, as with the forests, most of the world remains green.

In the face of all this, some obvious questions remain: 'Why, then, is the world green? Why do these plant-eating animals not devour *all* the available plants? And on the relatively infrequent occasions when they do, what has changed so that this can happen?'

The only exception to this picture of general greenness interposed with rare bouts of near-total destruction is when we look at our agricultural and horticultural plantings. Here the situation seems quite different – and much worse. On a regular, indeed constant basis, in all parts of the world, many insects are eating the plants we cultivate for our own use, and in such numbers that they can destroy our crops very quickly. To prevent this happening we must kill the insects first – and keep on doing so – otherwise this multitude of pests would leave precious little for our use. Nevertheless, in spite of our best efforts, each year they consume a significant proportion of our crops before we can harvest them; and continue their depredations whenever we store such produce for future use. These herbivores appear to be behaving like the locusts and budworms during outbreaks, and quite differently from most herbivores in nature. Presumably, too, their ancestors did not behave in this way when feeding upon the wild ancestors of our cultivated plants. Again we must ask: 'What has changed so that this can happen?'

Nobody would dispute that the world is green. Apart from the driest of deserts, or permanent icefields, plants grow on and cover nearly all surfaces of the Earth. They make up some 99.9 per cent of the weight of living things on Earth: only a tiny fraction of life consists of animals. If plants are removed from an area – by anything from fire to a bulldozer – they will quickly recolonise. Witness how soon plants start to grow on volcanic lava flows, or the tenacity with which they invade old buildings and other human constructions, like roads, once we cease to protect and maintain them. Think of some of the ancient cities found buried deep in the jungles of Central and South America. Nor are plants confined to dry land. Myriads of plants, from single-celled plankton to seagrasses and huge kelp, will thrive in water wherever sufficient light penetrates to enable them to photosynthesise.

Why then does the combined impact of all the many animals that feed on plants not make any impression on their number and volume? Why is it that, with the exceptions noted, herbivores seem able to eat no more than a tiny fraction of the huge amount of food that is there for the taking?

The usual answer to this apparent paradox is that before they can eat very much most plant-eating animals are killed by their natural enemies – their predators, parasites and diseases – or they kill each other competing among themselves for access to food or some other resource in short supply. Much of today's research by ecologists, and all of our attempts at biological control of pests of our crops, are based on the first presumption.

When we look closely at this explanation we find that some pretty involved logic has been used to arrive at it. Basically the argument goes like this:

- The great bulk of plants not eaten by herbivores dies and is quickly devoured by decomposers (mostly micro-organisms). These microbes

must be limited by their food because they eat all the dead plants. If they did not, dead plants would accumulate and form fossil fuels.
- Plants, too, must be limited by a shortage of their food (nutrients dissolved in soil water) because hardly any of them are eaten by the herbivores, yet they do not increase without limit.
- Herbivores, however, cannot be limited by their food because they eat so little of it. Nor is there any evidence that weather directly controls the numbers of herbivores; so this leaves only predators, or competition among themselves, to keep their numbers in check. On the rare occasions when they do eat most of their food plants, it must be because they have been 'protected' from their predators by human activity or 'natural events'.
- Finally it follows that because predators are regulating the numbers of their prey, they must, by their own actions, be limiting the amount of their food.
- The conclusion, then, is that all plants and all animals – except herbivores – are short of food, so their numbers cannot expand beyond the limits set by that food. Only herbivores are regulated by their predators, or by competing among themselves for limited resources, at densities below that which their food could support. So the world stays green!

But wait a minute. Why should herbivores be the exception? Might there not be an alternative and simpler – more parsimonious – explanation (a rigorous requirement of all scientific explanations)? What if green plants are *not* really the good food they seem to us? What if most plants are so nutritionally poor that, even in the absence of predators and competitors, most herbivores starve while eating their fill? Then all this deductive reasoning falls down, and we are left with the proposition that *all* animals, whether they eat plants or other animals, are limited by their food.

What evidence is there to support this proposition?

First we should ask: 'In what ways might plants, while remaining everywhere abundant, be an inadequate source of food for herbivores?' There are three ways they could do this: they might become too hard to find; become distasteful or poisonous; or become nutritionally so poor that few can survive by eating them.

Finding food is too hard

In the first case, the sort of plants that a particular herbivore can eat could be perfectly palatable and nutritious food, but so scattered and rare that the chance of the herbivore finding one among many other inedible plants is remote. To our eyes many species of plants are widely scattered and hard to find among other sorts of plants. However, this strategy of the plants has been readily countered by herbivores. They have evolved the ability to disperse with great efficiency, and to find their food plants no matter how infrequent and cryptic they may be.

This is particularly well illustrated by many small invertebrates like aphids and mites. Their bodies are so tiny that they will float away on the merest breeze. Many have evolved the behaviour of climbing to the top of a plant and launching themselves from it early in the morning. The air is warming and rising then, so they are quickly carried upwards and may travel great distances before falling from the sky late in the day as the air cools. (Try sitting out in the garden some cool summer evening after a hot day. Before long you will find small winged beasts landing on your clothes and starting to walk about.) For the great majority of these creatures the consequences of dispersing like this are grim. Most individuals will land where there is no suitable food plant, and quickly perish. For the population as a whole, however, the outcome is good; such great numbers are spread, and they become so widely scattered, that every suitable plant will be found. A lucky few of the many will land on the right plant.

I once witnessed a particularly arresting example of this power to find a rare host. I was walking across a large area of recently cleared and ploughed land and came upon a single 25 cm twig of *Eucalyptus* that had sprouted from a surviving root, and bore but half a dozen leaves. It was hundreds of metres away from any other green thing, and several kilometres from the nearest tree of the same species – a tiny target in the midst of a sea of bare earth. Yet on these few leaves were several 2 mm-long winged females of an insect that will feed on no other species of eucalypt. They were busily laying eggs. Later I was able to observe such females launching themselves into the morning breeze from the plants where they had grown, and catch some of them with a net towed by a light aeroplane 300 m above the land.

When a plant is found in this way it is quickly colonised as the animals multiply, producing enough progeny to devour the plant. Sometimes they do this. Usually, however, very little of the plant is eaten. Why is this so?

Food tastes disgusting or is poisonous

A second line of defence open to plants is to produce noxious chemicals so that they are distasteful or poisonous to any animal attempting to eat them. Plants generate a bewildering array of these chemicals. Or they could produce thorns, thick cuticle or hard seed coats to protect themselves from attack. They do this too. Again, however, herbivores have easily evolved counters to these strategies. They detoxify, sequester or simply avoid ingesting such chemicals, and circumvent physical barriers. A good example of the first is the poison 1080, widely used in Australia against introduced rabbits, pigs, foxes, cats and wild dogs. For them it is a deadly poison without antidote. However, it is a natural constituent of some Western Australian plants, and native animals in Western Australia which eat these plants are immune to it. In the eastern states, where 1080 does not occur naturally, these same native animals are not immune.

Many insect herbivores are not only immune to harmful substances in their food plants – they have become addicted to them. They need them as cues before they will attack a plant. Cabbage white butterflies are like this. They will only lay eggs and their caterpillars will only feed upon brassica plants – cabbages, cauliflowers, brussels sprouts, etc. – which contain specific toxins; those which give these plants their characteristic 'mustard' taste. Others have gone a step further, and have incorporated toxic chemicals from their food plant into their own bodies to deter attacks by their predators. The wanderer butterfly does this. Its caterpillars accumulate alkaloids from the milkweed plants on which they feed and these make the body of the adult butterfly highly distasteful to any bird which attempts to eat it. Most learn to avoid them altogether. Those that do attack them quickly spit them out and then avoid others of the same kind.

One consequence of countering these deterrent chemicals is that the herbivores that are successful at doing so are usually – like the butterflies mentioned above – specialists, each feeding on only one species of plant. So the plant has been successful in limiting the number of species which are able to use it as food, but not the number of individuals of an adapted species.

So having, by whichever means, neutralised this second ploy of the plants, adapted herbivores have the potential to eat out their food plants. But mostly they do not. Why not?

Food is not nutritious enough

The third way plants might avoid attack, even though they are abundant, easy to find, palatable and non-toxic, is to simply be inadequate food for the herbivores. They would do this if they lacked any one nutrient that animals must have in order to grow and breed. What is more, no animal could evolve a counter-stratagem to the absence of an essential nutrient. However, the common biochemistry of life precludes a plant from doing this; the chemicals needed to grow a plant are the same as those needed to grow an animal.

On the other hand, a plant could evolve to the point where an essential nutrient in its tissues is so dilute that a herbivore could not eat enough of the plant before perishing from malnutrition. Alternatively (because plants, too, must deliver nutrients into their new growth and their reproductive organs) the plant could limit the time that an essential nutrient is concentrated in its growing tissues or flowers and fruit. Then, while a herbivore may thrive by eating those tissues, it will be able to do so for only a short time. Soon it would again be reduced to consuming poor quality food.

As I shall discuss in this book, there is widespread evidence that plants have evolved both of these latter strategies.

Not surprisingly, then, we find that herbivores have, as they have with the other tactics, evolved a whole suite of structural, physiological, behavioural and life history adaptations to counter this dilution of their food. Nevertheless, once again, they rarely eat very much of the available plants. Why not?

Because, in spite of these adaptations, the third strategy has been relatively successful; for most of the time herbivores do not get enough good food. Specifically, their young seldom get food of sufficient quality to enable them to survive, let alone to grow. Those few that do grow to adults can then usually, but not always, get enough to maintain themselves. Only rarely and spasmodically, however, is their food nutritious enough, for long enough, to allow them to breed, and their new offspring to grow.

It would seem then, that if you are a herbivore, you can evolve ways to find plants trying to hide from you and you can counter or avoid poisons they produce to deter or kill you. But, having done so, there is little more you can do if you are then confronted with not being able to get enough basic nutrients from your food, no matter how much of it you eat.

So, the answer to the question 'Why does the world stay green?' is not the most widely espoused, and apparently obvious one: 'Because most animals that eat plants are eaten by other animals before they can eat the plants, or are prevented from eating them by other animals also trying to eat the same plants.' Rather it is one which is not intuitively obvious: 'Most herbivores starve while eating their fill of plants which look (to us) to be perfectly good food, but are actually quite inadequate food.' A universal feature of the life of all herbivores which illustrates this, and which is in stark contrast to that of carnivores, is the time they spend eating, the volume of food they consume, and the consequent volume of faeces they produce. They spend the greater part of their lives eating, constantly processing large amounts of poor quality food in order to extract sufficient nutrition from it.

To be more specific, plants are poor food for herbivores because they are mostly carbohydrate, and contain insufficient nitrogen for the production and growth of young herbivores. Furthermore, it is not just any old nitrogen that is in short supply. It is the nitrogen in quickly and readily absorbed amino acids that are essential for building new body protein. These amino acids are so dilute in plants for most of the time that herbivores are constantly striving to get enough of them. As a result they can produce few viable young, and most of those they do produce soon starve. And they will die whether or not others of their own – or any other – kind are trying to eat the same food.

I have referred several times to the commonly held belief that animals do not outgrow their environment because they compete among themselves for limited resources and the successful ones kill their competitors, or exclude them from access to the resources, so that they die anyway. But this belief is not tenable. Why?

There is no doubting that competition is a reality in nature. It is constantly observable and all-pervasive. And in this world it could not be otherwise. Once the first entities on earth (presumably simple DNA-like chemical structures) reached a stage of complexity where they could use other, simpler, chemicals in the environment to build copies of themselves, competition became inevitable. Why? – because, sooner or later, the supply of the least abundant of those elements which are essential for the building of new 'bodies' would run out. Once that happened only those better than others at gaining access to this now limiting resource would be able to make any more copies of themselves. And in doing so they would prevent others from using the resource. The unsuccessful ones would eventually disintegrate – 'die' – or be dismantled – 'eaten' – by the survivors which could then use their prey's released chemicals to build more of their own structures.

Since that presumed time competition has been a major force driving the evolution of more and more complex organisms over billions of years. Only those inheriting some attribute that made them better competitors survived to pass on their genes – or precursor genes – via new copies of themselves.

Much could be said about the role of competition in today's populations, but here I need make only two points. First, yes, it is vitally important in moulding the way in which plants and animals continue to evolve, because it decides *which few* of the many attempting to use limited resources survive. Whenever there is not enough for all, only those best adapted to out-compete their conspecifics survive and breed. Second, and of major importance to what this book is all about, *no*, competition does *not* decide *how many* individuals in a population survive. That is decided by the supply of the resource in short supply. Whether there are 1000 or 20 competing, if there is enough for only 10, then only 10 will survive. Competition is a consequence not a cause.

But what about the predators?

This leaves us with the other factor said to be preventing herbivores eating all the plants: predation. Predators are believed to be such efficient regulators of their herbivorous prey that they keep their numbers below the level that the available food could support. Yet this is not so. They are themselves limited by a shortage of their food. But not because they reduce the number of herbivores by eating so many of them. Their capacity to produce and raise young is constrained by their inability to catch enough of what seems, superficially to us, an abundance of readily available prey.

What, you may ask, is the evidence for all this? How can I justify such sweeping statements?

The rest of this book is devoted to explaining some of the evidence. It tells about many varied and fascinating ways in which herbivores have evolved to improve their access to the limiting nitrogen in their food, and how their predators fail to live up to their reputation as efficient killers. It also describes how some forms of competition have evolved that not only do not reduce the numbers that survive, but *increase* them. They do this by the highly inequitable allocation of what resources are available to just a few, thus ensuring a more efficient use of those resources. And, finally, it relates how it is the weather which is ultimately responsible for how much food there is, and so for how many animals there are.

Nitrogen – the key limiting factor

I should first explain why it is nitrogen, and not some other essential chemical – or energy – in food that is the key limiting factor.

Organised life on Earth is based upon four elements: hydrogen, carbon, oxygen and nitrogen, and it is fuelled by energy from the sun.

Many biologists believe, and base their research on the assumption, that what limits the growth of organisms is the supply of energy that they can access – from photosynthesis for plants; from plants for herbivores; from other animals for carnivores. The supply of solar energy is, however, to all intents and purposes, continuous and unlimited. Yet only a very small fraction of it is ever incorporated into plants and animals; most of it is re-radiated back to space as heat. Much less than 10 per cent of the energy reaching the Earth is incorporated into plants by photosynthesis. Only about one-thousandth of that is converted to herbivores, and the loss continues as herbivores are converted to carnivores, and so on, until only the original chemicals are left. If energy were the first to be limiting, would so much go unused? And would the little that is trapped be so wantonly wasted? For example, the evolution of warm-blooded animals would have been a very improbable event had the energy needed for their thermoregulation been in short supply. Similarly the large investment in energy required for long-distance migration by many birds is unlikely to have evolved if it were hard to come by.

The supply of the four basic chemicals, on the other hand, is not unlimited. However, carbon, hydrogen and oxygen are all very abundant and readily available. There seems little prospect they could run out. Nitrogen is equally abundant – but 99.95 per cent of it is inert gas in the atmosphere, and so unavailable to plants and animals. The remaining 0.05 per cent of the nitrogen on earth is combined with other chemicals, but half of this is in inorganic form and essentially unavailable to animals. The other half of that half of 1 per cent of all the world's nitrogen is in organic form. But 95 per cent of that is present as dead material in litter and soil or (mostly) as particulate and dissolved matter in the oceans. So, in contrast to the other three essential components of living things, nitrogen is in *very* short supply. And what little is available tends to be thinly spread in the environment. There is a relative, rather than an absolute shortage of it. Not surprisingly then, it is most often the first essential nutrient to become limiting for the growth and reproduction of both plants and animals.

Because of the inherited biochemistry of all life, nitrogen is required as a nutrient second only to carbon. It is the key component of amino acids from which proteins are built. And no organism – plant, animal or microbe – can survive or grow without a supply of nitrogen for the synthesis of proteins. Carbon, on the other hand, is greatly in excess of nitrogen in all living tissues. The ratio of carbon to nitrogen in the amino acids basic to all life varies from 1:1 to 2:1, while at the other extreme, in woody tissues of plants, this ratio reaches 1000:1.

Plants, of course, are the primary producers. Only they can fix energy from the sun. Animals must eat plants (or other animals) to obtain the energy to fuel their metabolism. Equally importantly, plants alone can incorporate inorganic nitrogen from the environment into organic forms that animals can then use to build their body proteins.

Plants must obtain all their nitrogen in solution from the soil, and all agricultural practice (including the use of manufactured fertilisers) attests to its acute shortage. Nature also illustrates this for us. The little carnivorous sundew or venus flytrap plants grow in soils with too little nitrogen to support normal plant growth and reproduction. They can survive and reproduce in such habitats only because they have evolved the capacity to catch and digest insects, thus supplementing the otherwise limiting supply of nitrogen with animal protein. But even then they are struggling. Feed them with more insects than they can catch naturally and they grow bigger and produce more flowers and seeds than those plants left to get by on whatever they can catch for themselves. Feed them with artificial nitrogen fertiliser and they can grow and reproduce without access to insect prey.

It is not hard to see, then, why a lack of nitrogen looms largest for herbivores, why it must be of equal or greater concern to the animals that depend on the plants for their food. Plants absorb nitrogen as ammonium or nitrate. Animals cannot do this. They must have ready-made amino acids manufactured by the plants.

For a start, however, herbivores are confronted with a food composed largely of carbon. Plants have used the great surplus of carbon in their environment for structural purposes, husbanding their scarce nitrogen to make protoplasm. As a consequence, most of the body of a plant is built of cellulose and lignin, both carbon-based. Animals cannot digest these tissues. So what nitrogen there is in the food of a herbivore is either locked away within indigestible cell walls, or is thinly and unevenly spread through the body of the plant. At best they can eat pollen or seeds, getting a food containing about 7 per cent nitrogen. At worst a diet of wood or xylem sap yields as little as 0.1 per cent nitrogen. Growing leaves will provide about 5 per cent. Animal tissues comprise around 15 per cent nitrogen, so they are mostly starting from well behind the eight ball.

But this is not the end of it. Much of the limited nitrogen that is present in the food of herbivores is in complex structural forms that require the expenditure of time and energy to break them down into the amino acids which the animals' digestive systems can absorb. It is only when the plant is transporting nitrogen as soluble amino acids to and from growing, reproductive and storage tissues that it is readily available. And all this is exacerbated by the fact that animals need much more nitrogen than do plants. Their structural materials are based on protein not carbohydrate.

Then animals have a third problem. Not only is nitrogen scarce in their diet, with much of it requiring expenditure of considerable energy before it can be absorbed, they cannot use all that they do absorb. The metabolic chemistry of all animals is such that in the process of converting nitrogen into body tissues, some must be excreted as metabolic waste.

And as I said earlier, carnivores, too, suffer from a relative shortage of nitrogenous food – but in a different way. While every individual animal that carnivores can capture is a rich source of useable nitrogen, for most of the time they just cannot catch enough of them, often enough, to meet minimal requirements for reproduction and growth. While to our eyes there may seem an abundance of prey just waiting to be caught and eaten by the carnivores, this is not so. Mostly the only prey predators can catch are the very young, the very old, the sick, the wounded or the momentarily incautious or just plain unlucky. As a consequence it is failure to breed on the part of females and early death from starvation of most neonates that limits the numbers of carnivores, just as surely as it does for herbivores.

In summary, first it is plants that are struggling to gain access to enough of the scarce available nitrogen in this world to support their reproduction and growth. In turn, the animals that eat plants are similarly striving to get enough of it. Finally the carnivores which eat the herbivores are struggling to gain access to enough animal protein to support their breeding and the raising of their young. So both herbivores and their predators are struggling to survive in an environment that is passively hostile and inadequate.

How herbivores access nitrogen

I said that herbivores have evolved a huge range of adaptations to improve their access to the limited amount of useable nitrogen in their food. To survive – as individuals and as species – they have had to evolve to cope with what was aptly referred to by a wise old scientist in an earlier generation as 'this universal nitrogen hunger'. However, before I discuss in more detail some of these examples, let's first have a look, in general terms, at what form these adaptations might take. I can identify six ways.

1. Herbivores could selectively feed on those parts of the plants which are richest in amino acids and synchronise their breeding and the raising of their young with times when the plants provide the greatest amount and concentration of these.
2. They might increase the concentration of this soluble nitrogen and prolong, in various ways, the time it is available in the plants.
3. They could eat more food more quickly, and extract, absorb and digest the available amino acids in that food more efficiently.
4. They could enlist the help of micro-organisms to break down components of their food which they cannot digest, and produce essential amino acids they are unable to synthesise themselves. Then they could devour their microbial 'helpers'.
5. They might supplement the limited amount of nitrogen in their food plants by eating other animals.
6. They could apportion and concentrate the limited amount of good food in their environment to a selected few individuals at the expense of the many.

Many of the tactics incorporated in these strategies have in fact been adopted by both vertebrate and invertebrate herbivores, young and old, male and female. Yet, in spite of all these adaptations, the chances are still very slim that any one individual will get enough good food to survive for long. Most young animals die either shortly after conception or birth. And this is why animals produce so many young. They must produce what appears to be a wasteful surplus of offspring to make sure that enough lucky ones find enough food to survive and replace them. Else their species would become extinct.

As an aside here, I should perhaps point out that the usual belief is the reverse of this statement. Most biology students are taught, and most educated people accept, that the remorseless struggle for existence in nature follows because organisms produce too many offspring. If they all survived, numbers would increase exponentially and the world would quickly be flooded with them. So the young must struggle against each other to survive – and most don't. But rather the reverse is true. No organism produces too many offspring. All produce so many young simply because each individual must struggle for existence. Surviving on this earth is, and always has been, especially for the very young, a struggle – a chancy business. The huge 'surplus' of young that all organisms produce is the universal illustration of this. The capacity to produce so many young did not evolve to provide a struggle for existence as a vehicle for evolution. It evolved because the only populations which persist on earth are those which produce sufficient offspring to ensure that at least enough of these gain access to sufficient food to survive and replace their parents.

And as we shall see at the end of the book, it is this universal great capacity to reproduce which permits sudden and huge explosions in numbers of animals when changed conditions in the habitat alleviate the usually chronic shortage of food so that many more young survive and grow to maturity.

Furthermore, those that die need not have been actively killed by a predator or out-competed by others of their own or another species. Most die because they fail to ever gain a foothold. For most animals the 'struggle for existence' is not a tooth and claw business. It is a lonely struggle to live in an inadequate world. They die young, and their passing is solitary, passive and unnoticed.

Those best adapted to the habitat of the moment – or just plain lucky to be at the right place at the right time – survive. Those that, for whatever reason, do not gain access to enough resources to survive, die – they are selected against. Natural selection is not a matter of 'the survival of the fittest'. As a Dutch colleague of mine famously states, it is 'the non-survival of the non-fit'. This being so, many must be produced to ensure some survive.

In our modern Western societies this harsh reality of the death of most young is largely forgotten: we have virtually eliminated such deaths from our own population. But for our early ancestors – and even for those of only one hundred years ago – it was commonplace, as it still is today for many peoples of the developing world. In the natural world it is, and always has been, the universal rule.

The chapters that follow highlight some of the myriad ways within the six general strategies I listed, that herbivores have evolved to increase their access to enough nitrogen to enable them to produce sufficient viable young to persist on earth. These in turn constantly illustrate why, in spite of their best efforts, herbivores for most of the time just cannot eat enough plants to prevent the world from remaining green.

2

Herbivores are fussy eaters

If you take the trouble to look closely at just what a herbivore is eating – be it a sheep grazing in improved pasture or a caterpillar eating gum leaves – you will find that it is a very fussy feeder. It will be highly selective not just about what sort of plant it will eat, but at what stage of its growth it will eat it and what parts of the plant it will eat.

There are many, many herbivores, large and small, vertebrates and invertebrates, which browse or graze leaves. None of them, however, will eat the leaves of just any plant, and a great number of them are 'host-specific'; they will eat the leaves of only one species of plant. The common cabbage white butterfly is one; its caterpillars will eat nothing but brassicas – cabbages, cauliflowers, brussels sprouts, etc. Its butterflies and caterpillars are addicted, as we noted before, to specific chemicals produced by this family of plants. But even those generalist feeders which eat many different species of plants will, when given a choice, select some species ahead of others: forbs ahead of grasses, legumes ahead of most other plants.

Seeking out the best: flush-feeders

Beyond this, however, whether they are host-specific or generalist feeders, nearly all of them will eat only the new growing leaves of their food plants. And this is equally true of mammals that eat grass as it is of caterpillars that eat pine needles, and of insects that suck the sap of plants. They are all what I call flush-feeders.

Koalas are true herbivores; they eat nothing but leaves. And they are host-specific; they will eat only gum leaves. Most people know this, and that, as well, they are very particular about which species of eucalypt they will feed on. Few realise, however, that in addition to being picky about what sort of gum tree they will accept, they are also flush-feeders. They browse through the crowns of trees eating nothing but the soft new growth. It is not that they cannot chew the tougher mature leaves; they can, and do, if there is nothing else to be had and they are very hungry. If, however, they cannot get a constant supply of new young eucalyptus leaves they will not breed, they will lose condition – and ultimately they will starve. In a bad winter, when there is little new growth on the gums, it is not uncommon to find dead emaciated koalas with their stomachs packed full of old leaves. The reason for this is that once a leaf is fully grown it contains much less nitrogen than it did when it was young, and the small amount it does contain is no longer present in easily absorbed soluble form, but bound up in largely indigestible proteins, and encased in tough, indigestible cellulose. Koalas cannot extract enough nitrogen from such old leaves even to maintain their body weight. So on a diet of nothing else they will waste away and die.

To grow, and to breed, they must have access to a concentrated source of soluble amino acids with which to build body protein. And their best supply of these is in fast-growing new leaves. Accordingly it is only when there is an abundant supply of these leaves that they can produce young and those young will grow to maturity. Furthermore the koalas' preference for the leaves of only one or a few species of gum tree is not capricious. They select those species that have the highest concentration of amino acids in their growing leaves.

The same selective feeding on the growing tissues of plants is seen with insects that suck the sap of plants. If you look at roses in your garden you will see that the aphids that are attacking them – another host-specific species – are all crowded just behind the tips of the soft growing stems and developing flower buds. And if you look carefully you will see that they are giving birth, almost continuously, producing young at a great rate. Once a stem stops growing, however, or a flower is about to expand, they quickly desert it, because from then on only water is being imported. When all growth on a rose plant ceases, most aphids die and the few that find enough food to survive cannot breed.

The examples of the rose aphid and the koala show that while becoming specialised to feed on new growth is all very well, a big problem remains. The new growth of any plant is mostly short-lived; nearly all growth happens in short spurts. Time in which to produce a new generation is strictly confined. So it is not surprising that, in addition to great fussiness about what they eat, the life histories of flush-feeders are geared so that their gestation and the early growth of their young are synchronised with times when these flushes of high quality food are present – when plants are actively growing, flowering and setting seed.

It is often stated that the reason for herbivores eating young leaves is to avoid the increased toughness and increased level of deterrent chemicals like tannins that accumulate in mature leaves. But, as I explained in the previous chapter, mostly they can cope with these if they have to. Usually they simply avoid the parts of the plants that contain these substances. Rather, they eat young leaves because the readily absorbed nitrogen in young leaves is quickly converted to insoluble protein in the mature leaves. It is then far less available, requiring much greater expenditure of energy to break it back down to a soluble form that can be absorbed and digested.

Yet even then selectively eating young leaves may not be enough – especially in a poor year when there is little new growth. At such times the females of many animals may resorb eggs or embryos, and the bulk of the few young that are born or hatched soon starve.

But concentrated amino acids are not only transported to growing leaves. They must also be delivered to developing flowers and setting seeds. And that's not all. When seeds germinate, the nutrients stored in them must be mobilised for transport to the growing seedling. Then, at the other end of the spectrum, a plant's leaves eventually senesce and die. Before that happens the nutrients in them must be retrieved and transported to storage organs like tubers or corms. From there they will eventually be shipped out again to sites of renewed growth. In all these cases insoluble proteins are broken down to amino acids that can be transported in the sap, and then converted back to proteins on arrival at the site of new growth or storage.

As you might expect, there are herbivores that have evolved feeding strategies that enable them to take advantage of all such movement of nitrogen in a form which they can quickly digest and put towards building their own growing bodies.

Let's look first at examples of animals using the flow of nutrients in and out of seeds.

Going with the flow: seed-eaters

A few years ago I was standing with a farmer in his paddock of barley while he told me about the problem he had with sulphur-crested cockatoos attacking his crop. As soon as the seeds start to germinate the cockatoos move in. They walk down a row of emerging seedlings, pull them out of the ground, eat just the base of the new stem, and then discard them. Once the new leaves are well developed, however, they leave the remaining plants untouched and depart. And they do not return to the paddock until the plants are fully grown, and the grain in the seed heads is starting to fill out. Then they descend again, push plants over, and eat the newly forming 'milk-ripe' seeds in the heads. But as soon as the crop is mature, and seed is set, they again depart, leaving the remaining plants untouched. 'Why would they do this?' he asked.

I told him that eating unripe seeds is common behaviour by these birds, although not often noted. I have watched white cockatoos attacking a large field of daffodils, picking the flowers off, carefully opening them and eating just the tiny soft developing seeds within the ovary. And each year I have to endure the same white cockatoos descending on my walnut tree, biting open the unripe green fruit and eating the soft immature nuts. The cockatoos, I explained, are 'homing in' on the soluble amino acids flowing into the new seeds before they are converted to relatively indigestible protein.

And they are not alone in this preference. The females of very many birds that subsist as adults wholly or in part on mature seeds turn to eating soft, milk-ripe ones before they lay their eggs; and they feed them exclusively to their nestlings. Another Australian parrot, the galah, in the wheat belt of Western Australia does this, concentrating first on the new seeds of weeds growing around the crop, and then moving into the crop as the seed-heads begin to form and swell.

Further evidence of the predilection of Australian parrots for the unripe seed of introduced plants is common. About the same time that the white cockatoos are attacking my walnuts, groups of their cousins, the yellow-tailed black cockatoo, are descending upon the many introduced pine trees growing around Adelaide, tearing open their green cones and devouring the developing seed within them. And many years ago in Wagga Wagga, the local galahs found my almond tree in the backyard. From then on they descended upon it each year, biting open the green fruit and eating the soft immature nuts within, leaving none to ripen for me. My neighbour here in Adelaide has an apple tree in his garden and each year the local rosellas attack the apples when they are still green, carefully discarding the flesh and eating just the white unripe seeds in the core.

But of course, this preference is not confined to parrots. The European goldfinch is another example. When females are maturing their eggs they eat an exclusive diet of milk-ripe grass seeds. And they also feed them, partly digested and regurgitated, to their young nestlings.

However, the best-studied example is the Australian zebra finch. In the wild these birds breed whenever there is an abundance of ripening grass seeds in their habitat; often several times a year, no matter the time of year nor the weather experienced. Yet at other times, no matter how much ripe seed there may be to eat, they do not breed. These pretty little birds have become popular as cage birds, but more importantly have become the ornithological equivalent of laboratory white mice for much experimental work. As a result, a good deal has been learnt about their physiology and nutrition. Experimental feeding has demonstrated that they will, in fact, breed only when they have access to milk-ripe seeds. Mature seeds are fine for maintenance of adults, but not for females laying eggs nor for their rapidly growing nestlings. Furthermore, it has been found that the reason for this selective feeding is that the ripening seeds, unlike mature ones, are equivalent to whole-egg protein, the essential nutrient for the development of embryonic birds. They

Figure 2.1 Zebra finches can maintain themselves on a diet of nothing but seeds, but can only breed when they have access to good supplies of ripening seeds. It is only at this stage that seeds contain the amino acids essential for the production and growth of young. Photo courtesy of Rob Drummond.

are a concentrated source of amino acids, including some nutritionally essential ones, which are not present, or present only in much lower concentrations, in mature seeds. This is the form in which nitrogen must be transported in the sap for storage as protein in the seed. Without access to nitrogen concentrated in this way, and in a form which is easily and quickly absorbed and digested, these birds cannot get enough protein to support the production and growth of a new generation.

So it is not too surprising to find that eating milk-ripe seeds is not confined to birds. It is widespread in the animal kingdom. Many mammals that mostly eat seeds, such as squirrels, will select immature ones whenever possible. Many that eat large quantities of fruit, like gorillas, prefer unripe fruit. Strange to our tastes, but this way they are getting the soluble amino acids being imported into the fruit and seeds – much more essential than the sugars in ripe fruit. And they will carefully extract and eat the immature seeds; ripe seeds are discarded or pass through them unaltered.

Codling moths attacking apples are another good example. They lay their eggs in the maturing flowers and when the grubs hatch they survive only if they can eat the developing seed in the newly developing fruit. The large tunnel in the flesh of the apple that so mars the fruit is made by the mature caterpillar eating its way out of the fruit to pupate.

The females of many bugs that suck the contents out of seeds cannot mature their eggs if they cannot feed upon soft developing seeds. There is one fascinating case where the grub of a fly bores into a seed just after fertilisation, eats the embryo and endosperm – but without killing the seed – and then diverts to itself the flow of nutrients originally destined to form the food reserve of the seed.

Readers who are gardeners will be as familiar as I am with the caterpillars that attack their green beans and unripe tomatoes. I'm talking about those fat green grubs that bore holes in the side of the fruit and then clean out the soft developing seeds within.

But remember the cockatoos in the farmer's paddock first attacked the germinating barley seedlings. I need to turn to a study of the ecology of a mammal to explain why they did this. Feral house mice living in and around irrigated rice fields in NSW subsist for most of the year on the large amounts of ripe grain spilt during harvest. But they do not breed. It is only when they can eat the ripening seeds of, first, the early weeds that grow around the fields once irrigation starts, and then, a month later, the ripening grain crop itself that they start to breed. Once the grain crop is mature, however, they again cease to breed, even though there is a great abundance of mature seed to eat. They are accessing the same high quality food as the cockatoos and the zebra finches.

In the laboratory the mice can be induced to breed again by feeding them on ripening seeds. But, as well, they will start to breed if they are fed on germinating rice grains. And in the field when there has been enough rain to cause the grain spilt on the ground to germinate, they will again begin breeding. They are 'homing in' on soluble amino acids being exported from the seed to the new growing seedling. The white cockatoos are doing the same thing when they pull up the young barley plants and eat just the growing meristem where the base of the shoot emerges from the germinating seed.

Prolonging the supply: grazers and gall-makers

One tactic that can improve the situation for a hungry herbivore is to induce plants to keep growing for a bit longer, thus prolonging the supply of good food. So long as conditions are still favourable for growth, most plants which are cut back have the capacity to put on a spurt of new growth, sometimes repeatedly. Many sorts of flush-feeding animals have capitalised on this. They graze the same plants in one place over and over again, so that – just like mowing your lawn – they keep producing a permanent sward of lush regrowth long after ungrazed plants around them have ceased to grow. The giant tortoises on the Aldabra Atoll in the Indian Ocean provide an excellent example of this behaviour. During the wet season they feed exclusively on 'tortoise turf', areas of mixed species of forbs and grasses which they graze repeatedly, maintaining all plants as new growth less than 1 cm high, and removing 90 per cent of the annual production. Their growing juveniles are even more selective, seeking out and eating just the rarer (and more nutritious) forbs within the turf. In the dry season when the turf stops growing, the tortoises are forced to eat much poorer food, browsing on shrubs and sedges – even fallen leaves. Then they turn to preferentially selecting parts of these plants with the greatest amount of soluble nitrogen – flowers, developing seeds and new growth. Their marine relatives, the green turtle, feed in vast areas of seagrass within which they similarly establish 'grazing plots', consistently re-cropping the same plants within these selected areas so that they live on an exclusive diet of young growing leaves.

Different species of wild geese, feeding on a variety of plants, in addition to being very selective of the species they will eat, graze as a flock, every few days harvesting maximum high-protein food from the same area.

Hares, red grouse and sheep on the Scottish moors eat the growing tips of heather shoots. They all have the ability to detect, and repeatedly browse upon, areas of heather as small as one square metre that have been fertilised with nitrogen. And they preferentially, and repeatedly graze areas of heather that put on flush new growth after they have been burnt (hence the management practice of continuous rotational burning of these heather moors to maintain red grouse populations for shooters).

Giraffes living in Tanzania's Serengeti National Park browse very selectively on the very young growth of the acacia trees. These new flush tips make up 80 per cent of their diet, and the animals produce persisting patches of this highly palatable regrowth by repeatedly grazing the same trees.

Limpets in the sea, and the larvae of caddis flies in freshwater streams, are animals that eat algae which they scrape off the surface of rocks. Both, in the same way as these other animals, repeatedly graze the same restricted (and fiercely defended!) area of rock, ensuring a constant supply of actively growing, high protein food.

Nor is this repeated grazing of whole plants the only way to prolong the production of high quality food. Insects that induce plants to form a gall achieve the same end, but in a more controlled and concentrated way. To form a gall, an insect must wound the actively dividing cells of a plant's new growth and inject a salivary secretion into them. This secretion not only causes the tissues where the insect is feeding to grow for longer than the surrounding tissues, but in a form different from that programmed by the plant's genes; they grow into the gall within which the insect lives. But creating somewhere to live is not the point of the exercise. If the insect within is a chewer it grazes the cells lining the cavity of the gall so that they continue to proliferate. If it is a sap-sucker it feeds on the contents of these lining cells. In either case the plant is stimulated to provide a continuing flow of nutrients into the cells the insect is feeding on. The gall-former has created a 'nutrient sink' delivering high quality food for much longer than would be the case if the insect just fed on the normally growing tissues.

Some argue that the true advantage of a gall is the protection it provides its occupant from attacks by its natural enemies. The evidence, however, is that most gall-dwellers are attacked by parasites and predators equally, or more often, than their free-living relatives. They are essentially sitting ducks, unable to escape from the galls which advertise their presence!

There are, on the other hand, many observations and experiments that support the nutritional explanation for galling. And the existence of what have been dubbed 'physiological galls' gives a clue as to how this way of life might have first evolved. There are species of aphids that settle on the bark of silver fir trees in Europe, and on balsam firs in Canada. They suck up the sap from living cells just beneath the bark. Their feeding stimulates these cells to enlarge, and in some cases proliferate, although there is no sign of any swelling on the surface. However, these cells contain much higher concentrations of amino acids than surrounding cells where the insects have not fed. Furthermore, newly hatched insects will preferentially settle at these sites and feed on these cells. And they grow faster and survive better feeding there than if experimentally forced to settle where the previous generation had not fed.

In the same way the European beech scale feeds at a single site under the bark of European and American beech trees. There it stimulates the cells in which it feeds to proliferate sufficiently to form a distinct zone of tissue dubbed an 'internal gall'. Once again these tissues have higher levels of soluble nitrogen, and newly hatched young scales seek out these sites to start their feeding. They are more likely to survive and they grow faster on these special galls than on undifferentiated cells.

There is another interesting example of feeding which stimulates regrowth but falls short of forming a gall. It is perhaps more akin to the repeated grazing by animals discussed above. In New Zealand there is a species of caterpillar that bores into the wood of living trees, making a large tunnel in which it lives, covering the entrance with a web of silk. It does not eat the wood, however, but chews the bark surrounding the opening to its tunnel. The bark responds with the well-known 'wound response' to produce a thick growth of callus tissue around the hole. The caterpillar repeatedly feeds upon this nutritious new growth, stimulating replenishment of the supply for as long as it remains in residence.

Creaming off the best: fast-track feeders

There are yet other ways that herbivores have evolved to increase their chances of gaining greater access to this precious resource of digestible nitrogen. One obvious one is to eat faster and/or spend more time eating. But even if an animal eats continuously, there are strict limits on how much food it can hold in its gut, and the speed with which it can digest the nutrients in that food, especially if it is of very poor quality (witness the case of the starving koalas related earlier). A number of diverse animals have got around this problem by what I call 'creaming off'; quickly extracting from their food just that portion of nitrogen that can be immediately absorbed. The rest, which would need to be held in the gut long enough to allow enzymes to break down cellulose cell walls and complex protein molecules before it could be absorbed, is discarded. On balance they gain more nitrogen this way than if they retained the food in the stomach for slow digestion of the recalcitrant portions.

If you feed brassica plants low in nitrogen to caterpillars of the white butterfly they respond by eating them more quickly; so much so that they grow as fast on these plants as on ones containing three times as much nitrogen. Technically they are feeding less efficiently because they assimilate less of the gross weight of the low-nitrogen plants they ingest. Nevertheless they win because they gain a higher proportion of the total nitrogen in that food. They pass several lots of food through their guts in the time it would take to digest all the nitrogen in one gutful, creaming off that which is immediately assimilable, and abandoning the rest. It is now known that this tactic is quite common among caterpillars of many species of moths and butterflies.

The giant pandas of China also employ this creaming off tactic. Most people know that pandas live on a virtually exclusive diet of bamboo. What is not so well known is that they have an alimentary tract that is not at all conducive to digesting a vegetarian diet. They evolved from carnivorous ancestors and have retained the simple gut of a carnivore without any out-pocketings and gut microbes. They partially overcome this handicap by very selectively eating only some parts of the bamboo; gaining four times more protein than from eating all of the plant. Yet this is not sufficient for their need for nitrogenous food. They have additionally evolved special teeth with which they can finely grind the bamboo, crushing more cells and making the contents available for quick assimilation. Yet they must still eat vast quantities (up to 6 per cent of their body weight each day), spend most of their time feeding, and pass the food through their guts in just eight hours, assimilating only the immediately available nitrogen.

Fruit-eating bats in Africa follow a similar regime. They chew up the fruit, spit out the fibre and swallow just the juice. This enables them to ingest two and a half times their body weight of juice in a night's feeding – and they pass it through their guts in a mere 20 minutes. Leaf-eating bats in Australia employ the same technique. They thoroughly chew each leaf, sucking out most of the liquid contents of its cells, and spit the fibrous remains on the ground.

As we saw, different species of geese repeatedly graze the same plants. But they also 'cream off' when eating the fresh grass. They can eat 25 per cent of their body weight each day, and pass the food through their gut in 30 minutes, defecating once every three minutes!

It used to be thought that some Australian brush-tongued lorikeets live almost exclusively on nectar, leaving a great puzzle as to how they could get enough protein from such a diet. Careful study of two Western Australian species showed, however, that their staple diet is not nectar at all, but pollen. Pollen is high in protein, but notoriously difficult to digest because the outer casing of the grains is very resistant to digestive enzymes. At best the birds would be able to gain access to the contents of something like 50 per cent of the grains they ate. However, these birds have very short intestines, and pass their food through very quickly. Retaining it any longer would not significantly increase the proportion of the pollen grains they could digest. Instead they absorb even more protein by passing as much pollen as possible through their system, as fast as possible, in the process creaming off that which is readily accessible.

In the case of all these animals this tactic inevitably involves trade-offs. The volume of food processed relative to body weight becomes huge. Far more carbohydrate – as sugars, starch or fibre – is ingested than can be used. Because of the low level of nitrogen in plant tissues all herbivores pass far greater volumes of faeces than equivalent sized carnivores, but for those going down this creaming off track the rate of production is even greater. Anybody who has parked their car under a tree infested with aphids feeding from the phloem sap of its leaves, rich in unneeded sugars, knows how much faeces – sticky honeydew – they can produce in a short time.

This massive passing of food through the gut reaches an extreme in sucking insects, called spittle bugs, that feed on the very dilute xylem sap which conducts water and dissolved inorganic nutrients from the soil via the roots to the rest of the plant. It can be 100 to 1000 times more dilute than phloem sap; more than 98 per cent water. As a result these insects may ingest 150 to 250 times their own body weight in 24 hours in order to gain enough food to survive and reproduce. Not surprisingly, therefore, they produce enormous volumes of faeces that commonly surrounds their bodies with a frothing mass of bubbles – hence their name.

Faced with such dauntingly dilute food these insects must cream not just some, but every last drop of nutriment from it. So it is no surprise to find that they preferentially feed on plants having the highest level of food – amino acids and amides – in their xylem. And the plants with the highest concentrations of these vital chemicals are those – like the legumes – that have nitrogen-fixing bacteria, housed in specialised root nodules, which pass organic nitrogen up the xylem. This preference has had one unexpected and economically expensive result. Vast areas of improved pastures in tropical America have been planted with introduced Old World grasses, mostly from Africa. All of these plants have nitrogen-fixing bacteria. These bacteria can double the level of amino acids in the xylem – usually not enough to produce any measurable increase in plant growth, but sufficient to significantly boost the nutrition of insects feeding on the xylem. As a result these pastures are subject to huge, ongoing outbreaks of otherwise relatively innocuous native American spittle bugs. Extensive plantations of sugar cane (another imported grass with nitrogen-fixing bacteria in its roots) growing from Mexico through to Brazil are similarly attacked.

Quite apart from this story illustrating the extremes that an animal can go to extract a living out of an inordinately dilute food, it reinforces the repeated theme of this book: how dependent herbivores are on the concentration of assimilable nitrogen available in their diet. And, I might add, as with the native caterpillars I witnessed attacking pine trees in New Zealand, nobody can claim that these insects became pests on these introduced plants because of the absence of their natural enemies.

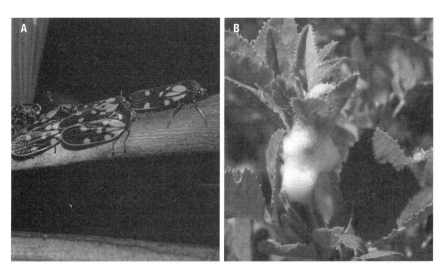

Figure 2.2   Passing enormous volumes of extremely dilute plant sap through their digestive systems in order to extract enough nutrients from it, means that these spittle bugs (A) produce vast quantities of watery faeces. They have evolved the capacity to form these into a protective froth which looks for all the world like a piece of spittle adhering to the plant (B).
Photos courtesy of Vinton Thompson.

So it would seem that most herbivorous animals feed selectively upon any part of a plant where there is an inflow of a rich supply of nutrients into actively dividing cells. They are flush-feeders. They do this because, just like the plants, they require access to this enhanced concentration of nitrogenous food to convert into protein to build new body tissue.

Catching the late run: senescence-feeders

As I said earlier, however, there comes a time when the flow is reversed; when nutrients are flowing out of a plant's tissues as they senesce. And there is another, much smaller group of animals, mostly sap-sucking insects, which has evolved to feed only on these tissues. Nevertheless they are doing the same thing as the flush-feeders. They are plugging in to a concentrated source of nutrients, rich in amino acids. But they are accessing these nutrients as they are being exported out of dying tissues. I call them senescence-feeders. Let me give you some examples.

Here in Australia there are many species of sap-sucking insects called psyllids, which feed on the leaves of different species of *Eucalyptus*. Some of these have evolved a very particular lifestyle. After emerging from the egg, they quickly settle and insert their mouthparts into the phloem to feed. However, unlike most sap-suckers, they will remain feeding at this same spot, growing through several nymphal stages until they fly away as mature adults some weeks or months later. Immediately they start to feed they begin building a cover over themselves. This is called a 'lerp', and the insect continually adds to it as it feeds and grows. The lerp is constructed from the insect's faeces. Like the honeydew of aphids, these faeces are a solution of almost pure carbohydrate left after the much scarcer amino acids in the plant's sap have been absorbed in the insect's gut. For both aphids and lerp insects, getting rid of this surplus carbohydrate entails almost constant defecation. Unlike the aphid's honeydew, however, a lerp insect's faeces solidify as the feeding nymph extrudes them from its anus, and it moulds them to form the lerp. This happens because the faeces of the lerp insect are starch, not sugar. Whereas other plant-eating animals have enzymes in their gut to break starch down to sugars that can then be absorbed, these lerp insects have evolved the capacity to do the reverse, linking surplus sugar molecules together to form starch.

Many lerp insects are flush-feeders, settling and living on new growing leaves. Interestingly their faeces are a mixture of insoluble starch, which forms the lerp, and sugary honeydew, which is discarded on the surface of the leaf. But one large group of species has evolved as senescence-feeders. They do not produce any honeydew. Often they live on the same species of eucalypt – even on the same individual tree – as a flush-feeding species. But they will not lay

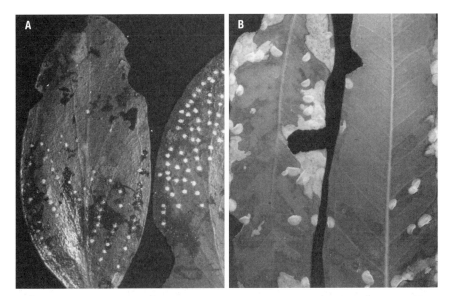

Figure 2.3 Australian lerp insects are sap-suckers. Some species (A) are flush-feeders and their young feed on nutrients flowing along a plant's main veins into a developing leaf. The dark areas are wet with honeydew. Other species (B) are senescence-feeders. Their young feed in the small ultimate veins of mature leaves where the breakdown products of senescence are first released. This species can further enrich the outflow of good food by making the tissues around its feeding site senesce more quickly than normal. This patch of leaf will redden (dark areas) and then die (white areas) soon after the adult insect has emerged and flown away, while the remainder of the leaf stays green and healthy. Photos by TCR White.

their eggs or feed on new growth, only on fully expanded mature gum leaves. And while the flush-feeding species feeds from the major leaf veins in which the nutrients are being shipped into the leaf, senescence-feeding individuals will feed only from the fine ultimate phloem elements situated in the lamina of the leaf between the veins. Why, I will explain in a minute.

The species that feed on flush growth can grow from egg to adult in about three weeks, whereas the senescence-feeders take about three months to complete their life cycle. This is not too surprising, as, in sharp contrast to the rapid inflow of nutrients to new growth, the mature leaves on which they feed take up to two years to die, very gradually releasing nutrients from their slowly senescing tissues.

To overcome this slow delivery of their food supply, these lerp insects have evolved an extra strategy. At the same time as they insert their mouthparts into the leaf to feed they also deposit salivary secretions into it. These secretions cause the cells immediately surrounding each insect's feeding site to break down and die more quickly. Over the weeks, as a young nymph steadily feeds and grows, an area of the leaf around it starts to go yellow, and then turns bright red. Finally, just after the adult insect has emerged and flown away, the red area dies, leaving a patch of dry brown tissue in an otherwise green and healthy leaf. And this is why the nymphs feed in the fine ultimate veinlets, not in the major veins. It is in these fine elements that the soluble nutrients released by the slowly dying tissues of the leaf first become available and are most concentrated. The action of the insect's saliva hastens the rate at which the leaf's cells are dying, thus further increasing the amount and concentration of amino acids in the sap it ingests.

There are other examples which highlight the difference, on the one host, between the fast-growing flush-feeder imbibing a strong flow of nutritious food being delivered to growing plant tissues, and the slower-growing senescence-feeder dependent upon a more gradual release of nutrients from dying tissues.

There are two species of scale insects which attack introduced ornamental ice plants in the United States. Morphologically they are almost identical, and they settle and feed on the same plant and feed from the same leaf veins. However, one species settles preferentially on young leaves, and completes two generations a year. The other settles on older leaves, has a growth rate of less than half of the former, and completes only one generation a year.

Two species of sawflies that mine the leaves of birch trees in the USA are another such 'pair'. Both lay their eggs on the same tree. But one species will select only soft expanding leaves. Its larvae grow very quickly and it completes several generations in a season. The other species lays eggs only in fully expanded mature leaves, grows very slowly, and completes but one generation each year.

The green spruce aphid, which feeds on the needles of spruce trees in Britain, is another senescence-feeder. But it has to move from leaf to leaf to track its food supply. Like the lerp insects, these aphids settle and feed only on mature needles. Each aphid induces the tissues surrounding its feeding site to senesce more quickly than the rest of the needle, producing a series of yellow bands that eventually coalesce so that the whole needle becomes chlorotic. In the autumn a few aphids colonise the mature needles which had flushed in the preceding spring. There they thrive, building up to high numbers. But when these needles, now 18 months old, die next summer, the new spring-flushed needles are still not acceptable. So the aphids' numbers plunge precipitously and the few that do survive lose weight and cease to produce young. As autumn approaches, however, the spring needles mature to the point where they are acceptable and the survivors from the crash in late summer move onto them and start a renewed build-up of their numbers.

A variation on this tactic of moving from leaf to leaf in order to track the availability of dying tissue is that of a leafhopper in Britain. It does not just move between leaves on the one plant, however, but between leaves on different species of plants. These leafhoppers complete their first generation each year on the mature leaves of evergreen blackberries formed in the previous growing season. There their feeding produces extensive yellow patches of premature senescence on these leaves. Nymphs will not move to the current season's leaves, even when severely crowded, and if experimentally placed on young leaves, will quickly migrate back to the old ones. However, the summer females, which these nymphs become, will not lay their eggs on blackberries. Instead they fly to the mature leaves of deciduous trees, usually oak, and lay their eggs there. The females that eventually arise from these second generation eggs then reject the tree leaves on which they were raised, and return to the now-mature current season's leaves of blackberries. There they lay overwintering eggs that will start the first generation again next spring.

Double-dipping

There are other sorts of senescence-feeders. Some are caterpillars, some are locusts; a few are vertebrates. But there are yet other animals that have evolved the ability to 'double-dip'; to take advantage of both growing and senescing tissues. By tracking both these sources of the flow of soluble amino acids in the plant they extend the time when they can gain access to a diet that will sustain breeding and growth. A good example of this ploy is that of the sycamore aphid in Britain. In spring, individual aphids grow and reproduce rapidly on the expanding new leaves of sycamore trees, and their population increases dramatically. In the summer, however, when the leaves are fully grown and the flow of nutrients into them has ceased, individual aphids still manage to extract enough food from a leaf to survive, but they cannot grow or produce young. But in autumn when the leaves start to senesce and export nutrients, the aphids once more resume growing and reproducing. The only exception to their enforced summer hiatus happens if there chances to be any leaves on a tree which are dying prematurely because they have been stressed or damaged. Adult summer aphids will quickly migrate to any such leaves and recommence breeding.

There is a leafhopper that feeds on rice in Japan that does the same thing. It feeds preferentially, first on new leaves that are still expanding, and then on old leaves which are becoming chlorotic. These are the two sites where soluble amino acids become most concentrated as they are, respectively, imported to growing tissues and exported from dying tissues.

Other insects which have adopted this double-dipping strategy are leafminers; those which feed on the internal tissues of a leaf while leaving the upper and lower epidermis intact. One such is a small caterpillar that mines the leaves of the evergreen oak in Israel. The female moths lay their eggs in freshly flushed leaves, and the new larvae eat nothing but the liquid contents of the cells in these new leaves. Once they are partially grown, however, and the leaves are mature, they start chewing up whole cells, enlarging the mine as they go. But the tissues of these leaves are now very low in nitrogen and it takes the caterpillars another ten and a half months to complete their development.

Perhaps the most extreme case of a double-dipping feeder is that of a small weevil which mines the leaves of hard beech in the North Island of New Zealand. In the spring the female weevils feed for two or three weeks on the expanding new leaves. So they are flush-feeders. By the time the leaves have matured and hardened, they start to lay their eggs in the mid-rib close to the base, one egg to a leaf. These leaves are shed by the tree within a few days of being attacked, but the eggs in them may not hatch for up to four weeks – not until the leaves are thoroughly dead. When they do hatch, the larvae complete their development in some three weeks, feeding entirely within the leaf lying on the forest floor.

We saw that gall-formers gain an advantage by inducing a plant to continue to import nutrients into their galls long after the rest of the plant has stopped growing. This means the insect inside the gall is able to go on feeding on high quality soluble nutrients long after the rest of the plant has ceased importing them. Some species of gall insects have prolonged their access to this good food even further. They do so by switching from this flush-feeding to senescence-feeding. Once their gall is fully formed and mature, they induce it to die more rapidly than the rest of the plant – much as the lerp insect does to a leaf – and in this way commence exporting nutrients stored in its tissues. While the insect is still growing, the gall will gradually discolour, eventually turn red, and then die and dry out just when the adult insect emerges to fly away and start a new generation. A species of phylloxera aphid which forms galls on the leaves of commercially grown pecan trees in the United States of America achieves this prolongation of good food, draining virtually all nutrients from an area of leaf around the gall before it cracks open to release the full-grown animal. Another sort of aphid that forms galls on the leaves of poplar trees in America has evolved a similar benefit in a slightly different way. The newly hatched young aphids settle on the still-expanding new leaves that then form galls wherein the aphids feed and grow. Soon the part of a leaf between a gall and its tip begins to yellow; if the gall is close to the petiole of the leaf the whole leaf will turn yellow. The aphids induce the leaves to advance their senescence from autumn to early summer, boosting the supply of nitrogenous food to the growing aphids before the gall dries and splits to release the adults.

There are still other insects, however, which have evolved a reversal of this double-dipping: they feed first on old tissues and then switch to new growth. In California the caterpillars of the checkerspot butterfly feed on deciduous plants that flush their new leaves in spring following the winter rains, and then shed them by mid-summer as the six month summer/autumn drought takes hold. The butterflies do not emerge until summer and lay their eggs on the now-mature leaves not long before they are shed. The hatching caterpillars feed and grow rapidly on these fast-deteriorating leaves. They are senescence-feeders. But they cannot complete their development before the leaves

Figure 2.4 The first time that this spruce budworm larva feeds in the spring, it must do so as a senescence-feeder, mining into year-old needles of balsam fir. Then, when the new needles start to expand and grow it will switch to them, completing its growth as a flush-feeder. Photo courtesy of Canadian Forest Service.

are shed. So they enter an obligate resting stage and do not emerge from this until the following spring when a new batch of leaves is sprouting. On these new leaves they resume feeding, now as flush-feeders, quickly completing their development and spinning their cocoons. From these chrysalids the new generation of butterflies emerges in the summer to repeat the cycle.

An equally fascinating example of this reverse double-dipping is that of the spruce budworm, famous for its huge outbreaks which destroy millions of hectares of balsam fir forests in Canada and the United States. Its caterpillars hatch from the eggs late in summer, and, without feeding, spin a silk 'parachute' on which they float away, dispersing far and wide. Those that are lucky enough to land on a fir tree immediately spin a web among the needles. There, still without feeding, they hibernate for the winter. In the following spring they emerge before the new fir needles have started to flush, and mine into and feed upon the old, now senescing needles of the previous year's growth. But as soon as the new buds start to swell and expand they move onto these and start eating the expanding new needles. They have switched from senescence-feeding to flush-feeding, and rapidly grow to maturity before the needles cease to import good food.

We can see, then, that while there are many different lifestyles, and many different parts of plants eaten, all the parts that herbivores select have the one thing in common: they are sites where soluble nutrients are most concentrated – and where the vital amino acids are to be found.

3

With a little help from microbes

Animals cannot produce the enzyme cellulase, so they are unable to digest the cellulose which, as we have seen, makes up the bulk of the tissues of plants. Nor, consequently, can they readily gain access to the nutritious contents of cellulose-walled cells of plants. So it is not surprising to find that many herbivores have evolved a variety of associations with micro-organisms like bacteria, fungi and protozoans, which *are* able to digest cellulose. Such associations markedly increase the proportion of a plant that a herbivore can use as food, and, in addition, converts the totally indigestible parts of the plants to protein-rich bodies of microbes which the herbivore can digest.

Dung-eaters

The eating of faeces – one's own or that of any other animal – is something that people consider to be quite disgusting, even harmful. In the animal world carnivores – and most omnivores – similarly will have nothing to do with their own faeces, usually burying them or depositing them well away from places where they eat and sleep. Not so for herbivores, however. Among a great many animals that eat plants, eating dung – their own or that of others – is a widespread and common practice with a perfectly sensible and adaptive function in this harsh and inhospitable world. It is technically called 'coprophagy'; literally 'dung-eating', from the Greek *kopros* = dung, *phagein* = to eat. Before I explain what its function is in nature, let me give you an example.

On tropical coral reefs at Palau in the Caroline Islands one can observe what I call 'cascading coprophagy'. Carnivorous fish, hunting on the reef, are followed by schools of herbivorous fish that catch and eat the carnivores' faeces as they fall through the water. Deeper down, other herbivorous species are harvesting the faeces of the herbivores above them. And so it goes, the high-protein droppings of the top carnivores passing down through a range of herbivorous species, providing a progressively poorer diet as it does so. Little if any, however, ever goes to waste on the floor of the sea.

Furthermore this is no casual event, but a repeatable and predictable practice. The herbivores actively seek out the carnivores. Some can consume 25 per cent of their own fresh weight in a two-hour feeding bout. Over 45 species may be involved. But none eats the faeces of their own species; only those of fish that eat a higher protein diet than themselves.

So, here is a behaviour that enables herbivorous fish to supplement their diet with a significant amount of extra protein. Without it their food of seagrasses or seaweed would barely support reproduction, let alone growth of their young. It ensures that more of the scarce protein in the habitat contributes to the next generation.

But this example is a little atypical. Most coprophagic animals eat their own faeces, or, when very young, that of their parents. Many adult herbivores actively feed their faeces to their young. Nevertheless, the ecological function remains clear – to conserve rare protein for the production and growth of young in a world in which most animals suffer a chronic shortage of protein. However, as I said, it is a practice nearly entirely confined to herbivores, and largely to those herbivores which are hind-gut fermenters. These are animals that have specialised micro-organisms living in the last part of their alimentary canal; the colon and its outpocketings, the caeca. There the micro-organisms multiply and are eventually passed out with the faeces.

These micro-organisms – usually bacteria, but also protozoans – digest the cellulose in the herbivores' diet; something the herbivores themselves cannot do. Many of these micro-organisms also use the unavoidably wasted metabolic nitrogen of their hosts: as urea in the saliva, or absorbed through the wall of the rumen of mammals; as urine recycled in the gut of birds and reptiles, or as uric acid in those of insects. In this way it is incorporated into the bodies of the microbes instead of being excreted in the urine or faeces. And some gut bacteria can fix atmospheric nitrogen. Finally, they all produce amino acids which the herbivore can absorb through the wall of the colon.

But the micro-organisms' bodies are themselves a concentrated source of digestible protein. However, as they live beyond the small intestine where the products of digestion are absorbed, this valuable source of protein would be lost when passed out in the herbivore's faeces. Hence dung-eating.

There is another large group of herbivorous animals which are fore-gut fermenters. Grazers mostly, like cows and kangaroos, which have micro-organisms in specialised outgrowths of their stomachs rather than their colons; at the beginning rather than the end of their alimentary tract. There the microbes do the same as those in hind-gut fermenters: digest otherwise indigestible cellulose and produce essential amino acids. But these sorts of herbivores will only eat their faeces if they are suffering acute protein malnutrition. Normally they do not need to do so because they digest their micro-organisms and absorb their manufactured amino acids when these pass back from the fore-gut into the small intestine. So a cow is not really a herbivore, but a 'cryptic carnivore'! She eats microbes that have first fed on grass taken into her rumen.

A particularly interesting fore-gut fermenter is the South American hoatzin. It is in many ways a most peculiar bird, and a strict herbivore. Like all strict herbivores it is very particular about what it will eat, feeding selectively on high protein buds, shoots and new leaves. In the breeding season females further restrict their diet to new growth of just four species of high-nitrogen legumes. Its crop and oesophagus have become greatly enlarged as the site of fermentation, but the crop remains highly muscular and has interior cornified ridges which finely grind the food. So the bird is essentially a ruminant, like the cow. Except that, unlike the cow, which must regurgitate its cud to chew it up, the hoatzin achieves both 'chewing' and fermentation at the same site! And an extract of its crop contents will digest cell wall material equally as efficiently as that from the rumen of a cow. However, it pays a price for enlisting microbes to supplement its low protein diet. Its sternum is much reduced to make way for the enlarged crop and oesophagus, reducing the area for attachment of flight muscles. As a consequence of this and the weight of its voluminous fore-gut, it is a very poor flyer, spending most of its time crawling through the branches of trees.

The hoatzin was thought to be the only bird that is a known fore-gut fermenter. But there is possibly at least one other, the kakapo, or night parrot of far-away New Zealand. Kakapo are very large (the world's heaviest parrot), flightless and solitary birds. Currently the species is staring extinction in the face but for the strenuous efforts being made to preserve and breed from the approximately 80 birds left alive.

They, too, are strict herbivores, even as chicks, living on leaves, grass seeds and fruit. They exhibit many of the special behaviours I have already described in other herbivores seeking to improve their access to soluble nitrogen. They select soft new tissues of leaves and chew up and extract just the juice from more fibrous tissues. They preferentially feed on pollen and unripe seeds. This latter preference extends to the same predilection for the unripe seeds of introduced walnut and pine trees as their distant Australian cousins. In addition kakapo have a specially ridged palate for grinding their (often tough and fibrous) food and the keel of their breastbone is very small, reducing attachment for wing muscles, and making way for an enlarged fore-gut.

Even with these special adaptations they mostly lack sufficient protein in their diet to be able to breed. The female raises her single chick on her own, feeding it on a concentrated mix of fruit and seed. This involves a tremendous and prolonged task of gathering food and bringing it back to the nest for the chick which rapidly grows bigger than her; while she wastes away to skin and bone. As a consequence, females attempt to breed only in 'mast' years –

Figure 3.1 The New Zealand kakapo is one of possibly only two birds that are fore-gut fermenters. Like the other one – the South American hoatzin – its enlarged fore-gut has resulted in reduction of attachments for wing muscles to the point where it can no longer fly. Photo courtesy of Don Merton.

seasons when their food plants set a great abundance of fruit and seed. Then they feed heavily on partially developed green fruit and seeds in the weeks leading up to breeding.

But to return to hind-gut fermenters and the eating of dung. When we look we find a wide range of plant-eating animals, from crustaceans to mammals, have this behaviour. And in all it serves the same function – supplementing a low-nitrogen diet of plant material with high quality protein.

Termites, or white ants, are a prime example. They live on a diet of wood. But not really. Like the cow, they are not the pure herbivores they appear to be. You will be hard-pressed to find food with less protein in it than wood; it is mostly cellulose and lignin which the termites are unable to digest. So they cannot survive on wood alone. They must enlist microbes to digest it, and then digest the microbes.

Termites harbour a large population of microbes, mostly protozoans, in their enlarged hind-gut. But these do more than digest the wood the termites eat; they also fix atmospheric nitrogen and recycle metabolic waste nitrogen. Experiments have shown that the rate at which they fix nitrogen can vary

200-fold, depending on how much nitrogen is already in the diet. But it can contribute up to half of all the nitrogen that a colony of termites needs.

As with all insects the nitrogenous waste of white ants is uric acid. It is excreted directly into the hind-gut where it is recycled by anaerobic micro-organisms which metabolise it and synthesise amino acids. This recycling can provide enough nitrogen to support up to 30 per cent of the biomass of a colony.

Termites probably can't digest the bodies of their microbes in the hind-gut to any extent, but they achieve this by actively eating their own faeces.

The indications are, however, that in spite of gut microbes providing a much enhanced and concentrated supply of protein food, this is often barely enough to get by. Whenever an opportunity offers, termites prefer to feed on wood that has already been attacked by fungi. And the more it is decayed – the more the wood has been converted to fungus – the more they thrive. Some white ants have taken this strategy much further. They farm a special sort of fungus, cultivating it in their nests on beds of their own faeces. These faeces are still largely plant material only slightly digested by gut microbes after one quick pass through their gut. After some six to eight weeks the termites re-ingest both faeces and fungi from the beds. But they also browse on the fungus as it grows on the beds. In particular they harvest special protein-rich spores which grow from the mat of fungus. As a final touch the winged adults, before they leave to establish new colonies, swallow several of these spores so that they can establish their own new farms.

One highly specialised example of the benefit of eating faeces loaded with micro-organisms is found in a species of Californian damp-wood termite. A reproductive pair starts a colony. They produce two sorts of faeces; ordinary dry ones and special ones made up of the bodies of the protozoans in the their hind-guts. They eat these special ones that contain 1400 times more protein than their ordinary faeces, which they do not eat. The females eat more of these special pellets than do the males. In fact a male gives most of his to his mate once she starts producing eggs. And when the eggs hatch both parents feed most of their protein-rich pellets to the young. A nice example not only of concentrating available protein to the production of eggs and the growth of young, but with father contributing to child-rearing to boot!

There is an American species of primitive cockroach, closely related to termites, which also lives entirely on wood, and carries microbes in special hind-gut pouches. They have taken this channelling of microbial protein towards the raising of their young a step further. The young of a pair of these cockroaches stay with their parents after they hatch, and feed on special anal fluid produced by the adults. This is a rich soup of concentrated microbes. The young quickly acquire micro-organisms of their own from this fluid, so that they could, at least in theory, then survive away from their parents. Yet they stay with them, continuing to grow while eating this high-protein diet, for more than three years! Perhaps they do this because life for them is still a knife-edge existence. Half of them die very soon after hatching, and only one-third survive their first year.

Studies with another cockroach more familiar to most, the ubiquitous German cockroach, have shown just how significant it can be for very young animals to gain access to protein via their parents' faeces. These insects are not, of course, herbivores. On the contrary they will eat anything organic. Yet their young, when first they hatch, are often restricted in their ability to forage for food. They can, however, survive and moult to the second stage (when they are much more able to forage) when prevented from eating anything other than the faeces of adult cockroaches. And they do even better if they are given only the faeces of female cockroaches.

Returning for a moment to fish, it was long thought that no herbivorous fish had evolved this sort of association with gut microbes, possibly because their juveniles, like those of all fish, are carnivores. But quite recently two species of buffalo bream, which are abundant in temperate and tropical waters of Australia, have been found to be true hind-gut fermenters. They both have greatly enlarged, thin-walled, caecum-like pouches which house many bacteria and protozoans. These fish eat red and brown algae, biting off large pieces and swallowing them whole. Unlike most herbivorous fish they do not chew their food up, nor do they have grinding mechanisms in the gut, to break up the cellulose walls of the algae. Instead they retain the food in the gut for a very long time (some 21 hours compared to the few hours for most herbivorous fish) allowing plenty of time for the micro-organisms to do the job for them. They usually feed in groups so dense that the water is cloudy with their faeces. Interestingly, when their juveniles first settle on the reef from their free-swimming life as carnivores, they swim among the adults. At this time they have fully developed caecal pouches, but with no microbes in them, and still eat small amounts of invertebrates. But they soon acquire microbes by eating the faeces of the adults, and quickly change to eating nothing but algae.

Many herbivorous mammals are hind-gut fermenters and are coprophagic, at least when they are young. The domestic horse is a good example. Like all domestic animals (and very few wild animals) its nutrition, and especially its protein nutrition, has been intensively studied. Foals from soon after birth start to eat their mother's faeces. This behaviour ensures that they acquire the necessary micro-organisms for their hind-gut. But they continue to be coprophagic, albeit at a decreasing rate, for six weeks. That this is a response to a need for extra protein in their diet is revealed by studies with adult horses, which are not usually coprophagic. If, however, they are kept on a diet which has just enough protein to maintain their body weight, they quickly revert to this behaviour, consuming all of their own faeces immediately after each defecation. Supplementing their diet with urea to raise its crude protein content stops the behaviour in as little as a week. Returned to the original diet the animals are again eating their own faeces within seven to ten days.

Herbivorous rodents like the voles, hares and rabbits, are hind-gut fermenters, carrying bacteria in their highly modified colons and specialised caeca. Their colons can separate liquid and bacteria from coarse particulate matter, and recycle it back through the caecum. And they all, like those dry-wood termites, produce two kinds of faecal pellets; soft and hard. The soft ones are pure caecal contents – concentrated microbial protein. The hard

Figure 3.2 The Australian ringtail possum is an animal which produces both soft faecal pellets of concentrated microbial protein and dry fibrous ones. Being nocturnal it eats the soft microbial pellets while resting during the day, and discards its hard dry pellets when moving about during the night. Photo courtesy of Bob Baldock.

ones are dry fibrous material containing little nitrogen. The animals selectively eat the soft caecal pellets, taking them directly from the anus when they are resting. They void their hard pellets when they are actively moving about.

Australian marsupials that eat foliage, like the ringtail possum and the koala, are also hind-gut fermenters. The ringtail exhibits all the gut specialisations of the rodents, and produces separate soft faecal pellets which it ingests while resting during the day. The koala on the other hand feeds these special pellets only to its young. As soon as the young koala puts its head out of the mother's pouch she starts feeding it on soft caecal faeces – called 'pap' – which she takes directly from her anus. She continues this supplementation of her milk for some six weeks during which time the youngster is growing exponentially.

The green iguanid lizards in Panama provide another example of the value of young animals eating the dung of their elders. They are the only known herbivorous lizards whose young are not carnivores. The young of all others start life eating invertebrates. Young iguanids are, however, from the moment they hatch, far more selective than older iguanids. They feed exclusively on the new growth of a legume that contains particularly high levels of soluble protein. And they pass this food through their gut very quickly, 'creaming off' the readily accessible nitrogen. Yet in spite of these adaptations, and the benefit of a reserve of yolk carried over from the egg, they must have the help of microbes, housed in their hind-gut, to succeed. Immediately the young iguanids hatch from the egg they climb into the foliage, where they seek out and closely follow adult iguanids. To obtain an inoculation of the specialised gut protozoans they need they must eat the faeces of the adults. One pellet would be sufficient for this, but they persist with this behaviour for three weeks or more, growing rapidly in the process. Artificially deprived of this continuing diet of microbial protein they hardly grow at all.

The various species of birds known as grouse that live in the northern climes of Britain, Europe and North America, are all herbivorous hind-gut fermenters. That is except when they are very young. Like the chicks of the domestic fowl, from the moment they hatch grouse chicks run free and fend for themselves, albeit aided and abetted by mum. And for the first weeks of their lives they eat mostly insects, gradually increasing the proportion of vegetable matter in their diet as they grow. And, like all herbivores, as adults they are very selective in what they will eat, concentrating on the buds of new soft growth.

The most famous of these birds is probably the Scottish red grouse, managed for centuries as a 'sporting' bird – i.e. for shooting. Successful management relies on their selective feeding. On the moors the heather (their sole food plant) is repeatedly burnt in a mosaic of small patches, maintaining a constant supply of new flush growth for them.

Figure 3.3 The Scottish red grouse is another hind-gut fermenter which produces both soft pellets containing concentrated bacteria from its caeca and hard, fibrous, low-protein pellets. Yet, unlike other hind-gut fermenters, there is no record of the adult birds or their young eating their highly nutritious caecal pellets. Photo courtesy of Andrew MacColl.

Like other hind-gut fermenters grouse all have a pair of large caeca containing bacteria. These bacteria break down cellulose cell walls to allow access to the cell contents, synthesise essential amino acids and recycle nitrogen from the breakdown of metabolic uric acid. The latter is transported as urine from the cloacal area (the common site of urination, defecation and egg-laying in birds and lizards) to the caeca. To do this the birds have evolved a retrograde flow in the intestine so that its contents move forward against the normal peristaltic flow back towards the rectum and cloaca.

Their caeca and intestine are much better developed and longer than those of their close omnivorous relatives, the pheasants, quail, partridges and turkeys, reflecting their much greater dependence on a diet of relatively low quality bulk food. And the smaller females have longer small intestines and caeca than the heavier males, better equipping them to cope with this poor food when developing their eggs.

Grouse, like their mammalian and marsupial counterparts, can separate small particles of food in the colon and concentrate them in the caeca, leaving the coarser material to pass out as faeces. However, they, like their mammalian counterparts, produce two types of faecal pellets, one made up of these coarse particles and the other of soft material derived directly from the caeca.

Strangely, however, unlike all other hind-gut fermenters they have never been observed to eat these caecal pellets. And I say 'observed' purposely.

These pellets contain more than twice as much nitrogen as the woody pellets. Even if the birds could absorb considerable amounts of nutrients directly from the caeca, eating these pellets would constitute a significant boost of bacterial protein to their diet. Especially for the newly hatched chicks, which, presumably, must anyhow eat some of their mother's faeces to become inoculated with the caecal bacteria. So, too, for the breeding hens, improving the quality of their eggs. It seems strange that such a rich source of readily digested protein should go to waste.

Furthermore, there are very few caecal pellets among many woody ones on the mounds of droppings under red grouse night perches; caecal faeces make up only 12 per cent of the birds' output of dry matter. Do the birds indeed eat much of their caecal droppings at night, but nobody has thought to look and see?

A final example. One showing a further curious twist to this adaptation of eating faeces; but without involving micro-organisms. The breeding hen of the European goldfinch eats only milk-ripe seeds, and feeds her nestlings on a regurgitated and partly digested paste of these called 'chyme'. At the same time, until the nestlings are about 10 days old, she eats their droppings and incorporates them into the chyme. Thereafter she discards the young's pellets over the side of the nest. In these first few days after hatching the young birds' digestive efficiency is minimal, so their faeces contain much unabsorbed protein as well as their metabolic nitrogen. So, feeding their droppings back to them during that time makes good sense: scarce protein, otherwise wasted, is recycled and used. The value of recycling this otherwise unused nitrogen to young birds until such time as they are able to fully digest it, is illustrated by being quite common behaviour in many species of birds, including those that are mandatory carnivores. The European swift, which feeds exclusively on insects, is one such. Like the herbivorous goldfinch, the hen eats all of the nestlings' faecal pellets for the first three weeks of their life.

All these examples are but a few of many, and all are variations on the same theme. Clearly, the eating of dung is common in nature. And in all cases it fulfils the same ecological function. It allows herbivorous animals to eat high-protein micro-organisms that have first digested the herbivores' diet of low-nitrogen plant material. In this way they gain the necessary minimum amount of protein needed for the production and growth of their young, and which they could not get from plant food alone. Occasionally we see that it can also provide a necessary protein supplement for the fast-growing young of carnivores.

Detritus-feeders

There are other sorts of animals that do not permanently harbour micro-organisms in their guts, yet still rely on them as an essential source of protein food. These are animals that have become specialised to eat dead plant material – detritus. And their mode of feeding indicates a way in which the more specialised use of internal microbes could have evolved. Some have internal microbes which break down detritus in their gut, but many rely upon external ones to digest the plant material. They may then eat both the detritus and the microbes growing on it, digesting the latter and passing the former out as faeces. Or they may graze just the microbes, leaving the detritus to grow another crop of micro-organisms. Similarly when the residual plant material is eaten it may be recycled a number of times, via coprophagy, each time being first enriched anew with fresh microbes. In fact 'detritus-feeder' is really a misnomer for these animals. They are actually living on a diet of microbes that have first grown on the dead plant material. It is not hard to imagine how ancient creatures like these could have evolved into the animals with highly specialised guts wherein equally highly specialised microbes live permanently.

The common Mediterranean woodlouse, or slater, has become widespread throughout the world, and is a familiar feature of most people's gardens. They mostly eat dead leaves, but also take live plant material and animal matter when available. If confined to a diet of pure detritus, however, they must become coprophagic, relying on eating their own faeces along with the detritus to subsist. But studies in England have shown that not just any old faeces will do. They have little interest in one-day-old ones, but avidly eat them when they are about three weeks old. At this age the faeces contain their greatest concentration of micro-organisms – mostly fungi – and 10 to 100 times more than uneaten leaf litter that has been aged for the same time under identical conditions! The scientists who studied this behaviour rather aptly named the process an 'external rumen'; the cultivation and then eating of micro-organisms is essentially the same as that achieved in the rumen of a cow, but before the microbes are ingested.

Another species of woodlouse found in Yorkshire has evolved a step ahead of the external rumen. These animals have a gut that functions in a way analogous to that of the mammalian rumen, but they do not have any obligate internal microbes. They depend entirely on ingesting those which grow freely on detritus in the field. Their hind-gut is nearly as long as their body, and is divided into distinct specialised regions. The first of these is a large sac in which the detritus is held for 24 hours while the microbes continue the digestion they had started before the woodlouse ate the detritus. The products of this digestion of cellulose are absorbed through the walls of the sac. The residual detritus, plus the micro-organisms, are then passed back to the next region. Here the microbes are digested by the woodlouses' enzymes and the products of this digestion absorbed through its highly papillated surface. The residual detritus is then passed back to the rectal region and ejected.

Not all detritus-feeders eat their faeces, however. Some, in fact, very specifically eat only the microbes growing on the detritus, as we saw earlier with some termites which cultivate specialised fungi growing on their faeces which consist of partially digested wood. There is another woodlouse that feeds on decaying sycamore leaves in Sweden which does a similar thing. A fungus grows on these leaves, gradually breaking down and digesting their cellulose. It grows into large black spots on the leaves – hence its name, tar spot fungus – and these colonies contain the highest levels of protein: pure, concentrated fungal tissue. The woodlice eat these colonies first, leaving large holes in the leaves. They will then gradually eat the tissues surrounding the spots, which contain much less fungus and are less nutritious. Only as a last resort will they eat the veins of the leaves which contain little nitrogen.

Turning to quite different animals in a different environment, there are small marine crustaceans called copepods that produce faecal pellets of fine particles of algae encased in a membrane secreted in their mid-guts. In the sea this skin is rapidly colonised by bacteria as they penetrate to the algal material within. The crustaceans then remove and eat just the membrane and its attached bacteria. The membrane is of little nutritive value but the bacteria amount to a significant supplement of high-protein food. An indirect benefit of ingesting just the membrane is that it causes the pellet to crumble and disintegrate in the surface waters where the copepods feed. Here bacteria again attack the algal particles, further increasing the supply of bacterial food for the copepods. And on the broader front this behaviour serves, as it does in other cases of marine herbivores eating dung, to salvage as much nitrogen as possible before it sinks beyond reach.

Many marine and freshwater snails are detritus-feeders. But, again, many of them are not, not really. They are actually grazers, eating nothing but the micro-organisms – diatoms, micro algae, bacteria – growing on the surface of living or dead plants. Others, however, are true coprophages. Two species of snail which live in the Thames estuary are a good example. They feed on the bottom deposits of organic debris, but are not evenly distributed through it. They are far more concentrated in places where the debris is fine-grained, and sparse where the deposits are coarse. There are many more bacteria in the fine deposits. The snails repeatedly ingest this fine debris and deposit their faeces back into it. In their guts they extract only the bacterial protein, returning the residue in their faeces to the bottom deposits where new bacteria grow on them.

A fairly extreme case of a herbivore gaining nutritional benefit from micro-organisms is that of marine shipworms. Anybody who has had anything to do with boats and the sea will know about these creatures and the damage they can do to any wood left in the sea for long. They are actually not a worm, but a very modified type of bivalve clam that bores into the wood and lines the tunnel with its shell. They are adapted to a way of life in many ways akin to that of termites on the land. As we saw, wood is a food with perhaps the least nitrogen of any.

The young worms are free-swimming and live by filtering plankton out of the water with their gills. They continue to do this after they have settled and built their tunnels. However, they have been experimentally grown to maturity in seawater filtered free of plankton, so can subsist without this source of animal protein. Probably they can get some of their protein by digesting the marine fungi which quickly infest any wood in the sea. Mostly, however, they are dependent on nitrogen supplementation by very specialised bacteria living in special glands in their gills.

These bacteria can digest cellulose, synthesise essential amino acids and fix nitrogen. In a species of shipworm which lives in the Sargasso Sea (where the density of plankton is notoriously low) these bacteria can fix nitrogen at a rate which doubles a worm's cellular nitrogen in one and a half days. This is more than 20 times faster than they do in several other species of worm living in coastal waters where there is an abundance of plankton. The bacteria in young worms can fix nitrogen even faster than those in adults. These young ones have been known to increase their length more than 30 times in a month. They could not filter plankton fast enough to sustain such a rate of growth.

Yet even with the aid of fungi and bacteria, on a diet of wood alone these worms still live on a hairline between sufficiency and deficiency of nitrogen. With plankton excluded from their diet many species of shipworm can grow but are unable to reproduce.

Finally there is a group of herbivorous animals that has really fine-tuned the business of enlisting micro-organisms to improve their nitrogen nutrition. These are insects, mostly aphids and psyllids, which feed on an exclusive diet of the phloem sap of plants. This sap is rich in sugars, but low in nitrogen. In fact it often completely lacks some amino acids that are essential for the animals to survive, let alone grow and reproduce. All these insects have a special organ outpocketing from the gut called a mycetome. Within this live a variety of specialised micro-organisms – bacteria, usually, but also in some species fungi or virus-like organisms. So specialised are these microbes that they have become physically simplified, can live only in these special mycetomes, and are passed from generation to generation via the insects' eggs. However, unlike the microbes in the cow's rumen, they are not digested by their host. Instead they produce essential amino acids which the host insect absorbs as a vital addition to its plant food.

4

Meat-eating vegetarians and cannibals

The tammar wallabies which live on Kangaroo Island in South Australia, were considered to be strict herbivores. However, I have a colleague who spent many hours quietly following them and recording what they ate. To his surprise – and the disbelief of many – he discovered that they are not the obligate vegetarians everyone assumed them to be. On the contrary they are frequently carnivorous.

They commonly eat mice and nestling birds, small lizards and insects whenever they encounter and can catch them. And they feast upon large hepialid and cossid moths when these are synchronously emerging from the soil. Beyond this opportunistic carnivory, however, they appear to be systematic hunters. All the time they are foraging and browsing they stop, cock their heads on one side, and listen. Then they dig rapidly into the litter and humus catching and eating insects there. What is more, it seems they are not alone in this behaviour.

A TV documentary, 'Rock Opera', presented by the Australian Broadcasting Corporation in 1997, has dramatic scenes of Queensland rock wallabies capturing and eating large sphinx moth caterpillars. These wallabies, too, had been considered, even by the scientist studying them, to be strict herbivores. He had never observed them to eat anything other than vegetable matter. It was pure chance that the film crew observed and decided to film this behaviour at a time when he was not present. And it has now also been discovered that insects are an important supplement of the diet of the rufous hare wallaby, especially during drier times when there is little browse about.

A further indication of the enthusiasm wallabies have for animal food is revealed by two scientists' accounts, one of a pet Queensland rock wallaby, the other a pet swamp wallaby, snatching cooked chicken from the table and eating it.

Strictly vegetarian?

But this phenomenon is not restricted to wallabies. A colleague of mine working with captive sugar gliders saw them attack and eat house mice entering their cage. Even more to her astonishment, she discovered them killing and selectively eating just the brains of nestling budgerigars which she had kept – she assumed safely – in the same cage.

Two common species of Australian desert rodents, the spinifex hopping mouse, and the sandy inland mouse, were generally thought to eat nothing but seeds. However, a thorough study of their diet and feeding behaviour found that both are omnivores; invertebrates comprise an important component of both their diets.

Another Australian mammal which was believed to be a strict vegetarian is the dugong. Not so it seems. These large marine beasts, like some other herbivores I have mentioned, are repeat grazers of seagrass, thus ensuring a constant supply of flush new growth, high in soluble nitrogen. However, recent research has revealed that they are omnivores. They feed extensively on animals called sea squirts, bursting their hard exterior and eating the soft contents. They have also been seen to eat fish caught in nets.

Australian fruit bats have been observed eating lerp insects, and captive individuals of two Old World species were found to show deliberate and innate behaviour in tracking, catching and eating insects entering their cages.

And so it goes. How many more cryptic meat-eating vegetarians do we have in Australia just waiting to be observed by somebody looking for this behaviour? Or how many such observations go unrecorded because they were thought to be aberrant? Like another colleague of mine who has seen brushtail possums eating road-killed animals, but never thought to record this until discussing the topic with me.

Looking further afield than Australia we find that there are many more supposed vegetarian mammals which supplement their diet with animal food. And in addition to predation on other species, their carnivory often takes the form of cannibalism, or scavenging on the bodies of dead animals.

Various North American squirrels commonly take insects from foliage and under bark, and preferentially select acorns that are infested with insect larvae to store in their winter caches. They are also general scavengers on dead animals and are active predators of a number of small vertebrates including young snowshoe hares, nestling birds and lizards. They have even been recorded eating soil soaked in human urine, such is their apparent hunger for nitrogen! And a colleague of mine in Italy has repeatedly observed European red squirrels feeding on adelgid galls on spruce trees. These galls amount to a solid ball of soft insect tissue.

Some voles are known to be omnivorous, but most are thought to be entirely vegetarian. This was so for two common European species. One was considered to be an 'extreme herbivore' feeding almost exclusively on grasses, and the other a specialist seed-eater. But intensive study of their diets revealed that in spring and early summer – when they are breeding – 10 and 30 per cent of their respective diets consists of soft-bodied immature insects.

In North America the prairie vole – another apparently strict herbivore – has been seen to gorge on periodic cicadas emerging from the soil in vast numbers in spring. As these insects emerge only once every 13 to 17 years, this would hardly constitute a reliable and regular diet for these short-lived little mammals. However, it does show that they will take animal food when the opportunity offers. It may well be that they, and many other microtines, most of which are presumed to be obligate vegetarians, turn to eating animal food much more commonly than is believed. Close observation, especially immediately before and when they are breeding, may well prove rewarding.

The insect- and meat-eating propensities and aggressive predation of monkeys by chimpanzees are now well known. But their (and our) near cousins the gorillas were considered to be totally vegetarian. Once again, however, more recent and thorough studies have shown that all is not what it seems. Insects, especially ants and termites, but also caterpillars, are an important and regular component of the food of gorillas, conservatively estimated to account for 5 per cent of their diet. As in so many cases where insects are eaten by seemingly herbivorous animals, it is mostly the immature stages which are taken. This means that there are few if any hard parts left in samples collected from their guts or from their faeces to indicate the presence of animal tissues in their diet. So it is with the gorillas. They break open the sides of termite nests and eat the soft-bodied workers within. They pluck the nests of weaver ants – about 5 grams of soft eggs, larvae and pupae enclosed within leaves bound together by adult workers – and eat these like sandwiches!

Many species of monkeys that have been thought to eat nothing but leaves or fruit, prove on closer study to have a keen interest in eating meat whenever the opportunity arises. African blue monkeys are one such group. They eat invertebrates, especially during the breeding season, and are also opportunistic, possibly regular predators of small animals, like flying squirrels, birds and lizards. Madagascan lemurs, too, are now known to eat both invertebrates and vertebrates. It is likely that meat-eating and regular predation on vertebrates, as well as the eating of immature invertebrates, is far more widespread and common among primates than currently recognised.

The most recent example of things not being what they seem is the discovery that the hippopotamus is not the obligate vegetarian that naturalists have always depicted it. While known to be aggressively territorial and to readily attack and kill other species of large animals, including humans, there were no records of carnivory associated with such attacks. Now they have been observed killing and eating impala entering the water, scavenging the prey of wild dogs and crocodiles (this latter was also depicted in David Attenborough's documentary on crocodiles), and of cannibalism.

A mammal that is normally entirely herbivorous is the giant panda of China. As I explained in Chapter 2, pandas belong to the large group called Carnivora, and have a simple gastrointestinal tract anatomically similar to that of other carnivores. Yet, as I reported, they live on virtually nothing but bamboos. However, they can survive on this food only by virtue of a number of behavioural, anatomical and physiological specialisations that enable them to extract maximum protein from this plant diet. But this is still at considerable cost. They cannot accumulate fat reserves on their poor quality diet and are minimally active for most of the time. And their capacity to reproduce is severely limited, with low rates of pregnancy and poor survival of the few young that are born. Perhaps not surprisingly, therefore, they retain an appetite for, and the capacity to assimilate, animal food whenever they can obtain it. Remains of monkey, rodent and musk deer have been found in panda faeces, they have been seen catching and eating rats, are strongly attracted to meat in traps, and will eat meat regularly when it is offered to them in captivity.

Even being a domesticated animal, with the benefits of protection from predators and diseases and an enhanced, balanced nutrition, is no guarantee of access to an ideal diet. Often, in spite of retaining the ability of their wild ancestors to selectively feed on plants with the greatest content of available nitrogen, domestic stock suffer from serious protein malnutrition. So it is not too surprising that there are many records of sheep and cattle – and of managed populations of deer – scavenging the flesh of dead vertebrates and gnawing at their bones. And both sheep and deer have been observed in Scotland to kill and eat chicks of seabirds nesting on the shore. But such reports have, to date, nearly always been attributed to the animals suffering from some form of mineral deficiency rather than to any intrinsic shortage of protein.

The list then, is diverse, and growing as more people are alerted to look for such behaviour. In the past it may not have been recorded simply because nobody thought to look – a herbivore is a herbivore, so there is no point looking to see if it eats animals. In other cases, it is probably because the composition of diets is most often deduced from samples of the contents of the alimentary canal, or of faeces. In such samples little or no trace of the soft animal tissues usually eaten is likely to remain (think of the chance of finding the internal tissues of the sea squirts eaten by the dugong).

All the above examples are of feeding by adult animals, although, significantly, it is breeding females rather than males that most avidly seek out animal food. A recent excellent illustration of this dichotomy between the sexes comes from a study of ring-neck ducks in North America. These ducks are normally largely vegetarian, but do eat some invertebrates. When they are breeding, however, the females greatly increase their intake of animal material, while males do not.

Of even greater significance, however, when we turn our attention to what *young* herbivores eat, we find that a diet of animal protein is virtually universal for them – and mandatory.

Starting out carnivorous

All mammals start life as carnivores, subsisting on nothing but pure animal protein (I like to tell dedicated vegetarians about this!). At first, as embryos, they are nourished by the body of their mother via the placenta. The better her nutrition, the more they will thrive. Then, beyond birth, they are dependent upon an exclusive diet of milk. Here, too, the quality of this food is dependent upon the protein nutrition of the mother both before and when she is nursing. Mammals that are truly vegetarian as adults can only be weaned on to a totally vegetarian diet once they are past their early period of very rapid growth. And then but gradually, weaning is not usually complete until the young animal is fully grown: indeed in many instances bigger and fatter than its mother!

Birds and reptiles are similarly nurtured initially by the high-protein yolk of an egg, supplied from the body of the female. Again, the quality of this food is dependent on the quality of her diet. Many newly hatched birds and reptiles continue to benefit from a yolk sac carried over from the egg. Such yolk sacs may sustain them for days or weeks without eating. This is also true for many fish and invertebrates.

But, lacking the mammals' advantage of an ongoing diet of milk, nearly all otherwise totally herbivorous birds and reptiles must eat some form of animal protein during their initial period of exponential growth.

The young of birds like grouse, capercailzie and ptarmigan, which eat nothing but leaves as adults, must eat insects for the first days after they hatch. The Scottish red grouse is a good example. As adults they eat nothing but the growing tips of one species of plant – heather. And they have evolved a range of specialised behaviours and physiology to cope with this diet. Nevertheless their young chicks, which are free ranging from the time they hatch, eat nothing but insects for the first three weeks after they hatch. Additionally their otherwise strictly heather-eating mothers eat a significant quantity of insects before they start to lay their eggs.

Another bird, far removed both geographically and taxonomically from the red grouse, that lives entirely on plants is *Notornis*, the New Zealand takahe. It was long thought to be extinct, and still lives a very tenuous existence in the inaccessible fiordlands of the south-west of the South Island. These birds spend all but the harshest winter months in the tussock grasslands above the tree line. Here they feed almost exclusively on tussock grass, selecting out just the actively growing basal two to three centimetres of the stems and discarding the rest of the plant. Each year females nest and lay eggs but rarely raise any young. And this seems to be because from the time they hatch, and for some time thereafter, their chicks must have an exclusive diet of insects. They are aided in this by their parents which scratch characteristic areas of some square metres clear of moss and litter to expose insects for the young to eat. These characteristic 'scrapes' are only found where and when there are young present, and it seems that it is only in seasons when insects are especially abundant that any chicks get enough of this animal protein to survive.

Various species of geese are, as we saw earlier, highly specialised grazers on flush plant growth, with a number of adaptations that enable them to cope with this potentially limiting diet. Yet in spite of these, goslings are obligate consumers of significant quantities of invertebrates in the first few weeks of their life. This is equally true for the young of otherwise totally herbivorous ducks.

The Australian mistletoe bird was another thought to be strictly vegetarian – but even more exclusive, eating only the fruit of the parasitic mistletoe plants. Yet, here again, close study reveals that they also include insects in their daily diet and feed their nestlings nothing else for the first few days after they hatch.

Then there are the hummingbirds of North America – quintessential herbivores, with an insatiable appetite for high-energy nectar. Few people realise, however, that these tiny dynamos must eat insects every day of their lives, and must feed their young hatchlings on nothing else but insects. And so it is for many other birds, including Australian honeyeaters, which had been thought to subsist on an exclusive diet of fruit or nectar. They all eat insects, especially when they are young.

The same picture emerges for a whole range of birds which feed exclusively on seeds. To varying degrees they must feed their nestlings on a diet of insects, or at least a mix of pre-digested seeds and insects.

Many Australian parrots are seed-eaters yet their breeding females commonly eat insects and feed their newly hatched young an exclusive diet of them. One particularly fascinating example is that of corellas in Western Australia. Females with young chicks dig up large insect grubs in the soil, split them open, feed the near-liquid contents to their nestlings, and discard the tough external skeleton. What chance is there of detecting the presence of such animal food in samples of the gut or faeces of these birds?

Another is that of rosellas in New South Wales eating communally living caterpillars that live on eucalypt foliage. To do this they first must tear open the 'tents' of woven silk and dead leaves in which the caterpillars live. They have even been observed teaching their young how to do this!

Among seed-eating birds the pigeons have a specialisation that illustrates the many quirky turns that evolution can take under pressure of the need to provide growing neonates with animal protein. These birds feed their young on 'crop milk'. This 'milk' is actually rapidly proliferating cells of the lining of the adult birds' crops that they regurgitate and feed to their nestlings. And it is a protein food as nutritious as milk or egg-yolk; and more readily digested. Nestlings fed on this milk can grow faster than any other young vertebrate, so pigeons are able to produce successive broods much more rapidly than other birds.

There are very few truly herbivorous reptiles, and those that are (like many iguanids, including the famous marine and terrestrial ones of the Galapagos Islands), are carnivorous or have access to microbial protein when they are very young.

Adult green turtles, as we have seen, eat seagrass, which they crop repeatedly to produce 'lawns' of new flush growth. However, they are not as exclusively herbivorous as they seem. In turtle farms they will readily eat fish, invertebrates and high-protein pellets. As a result they grow much faster and have a higher rate of reproduction than animals in the wild. Even there, however, they are not exclusively herbivorous, having been recorded eating invertebrates and fish. Furthermore, their young are entirely carnivorous. As soon as they hatch in the sand, they run down the beach and swim rapidly out to sea where they famously 'vanish' for their first year or so of life. It is only quite recently that we have learnt where they go and what they eat. They live in the Langmuir bands far out in the open ocean where they eat nothing but planktonic animals which concentrate in these bands.

There are some fish which are vegetarians – or supposedly so, most of them eat algae, a very different food, nutritionally, from vascular plants. However, some do eat aquatic vascular plants like the true seagrass. Nevertheless, none of these fish is vegetarian as juveniles. Once having passed the stage of pelagic larvae which hatch from the eggs (and usually have non-functional mouths and alimentary systems), and having absorbed their yolk sacs, they prey on small planktonic animals. But not all young fish are pelagic. Some are cared for and fed by their parents. Here an unusual form of 'cannibalism' is found in some cichlids. The adults produce copious amounts of mucus from greatly enlarged mucus cells in their skin. The newly hatched young feed on this mucus, constantly picking at their parents' flanks. It has proved to be impossible to raise such young fish in captivity if they do not have access to this parental mucus.

And let's not forget the world of the invertebrates. Here, too, many apparent herbivores are not what they seem. Especially is this true when we look at females generating progeny, and particularly at their rapidly growing young. Then we find widespread and persistent carnivory among putative herbivores.

Land crabs in different parts of the world were mostly considered to be strict vegetarians. Recent studies have revealed, however, that adult crabs eating only plants are usually living on a diet deficient in protein. So whenever the opportunity arises they are carnivorous. They will eat a variety of prey – small worms, insects, crustaceans (including their own juveniles) and in captivity, meat. When experimentally supplied with such protein supplements they not only grow faster and produce more young, but are much less prone to cannibalise their young. And, of course, their young, which hatch from the eggs in the sea, are free-swimming pelagic larvae that feed entirely on microorganisms in the plankton until they metamorphose into tiny crabs and come ashore – only to have to run the gauntlet of being eaten by their parents!

Another interesting example is that of vegetarian slugs eating insects that have become trapped on the sticky leaves of carnivorous plants (and, yes, as I recorded in Chapter 1, by supplementing their 'diet' with animal protein, these plants can survive and reproduce while growing in soils which lack sufficient nitrogen to support their reproduction).

But there are many other herbivorous molluscs that are occasional carnivores; often obligate ones when they are young. The European edible snail is a good example. In batches of eggs the first to hatch quickly start to eat their unhatched fellows. In some species of snails this propensity has become entrenched. The snails include 'trophic' or 'nurse' eggs in each batch of eggs they lay. These are infertile, larger and much more abundant than the fertile ones, and are routinely eaten by the young hatching from the latter.

And the same adaptation occurs among vertebrates. One recently discovered example is that of a species of frog which breeds in very small pools of water in such places as tree holes. The female deposits a single egg in the water where the male fertilises it. However, the water does not contain enough food to sustain the subsequent tadpole to maturity. So the pair return repeatedly to the pool, each time engaging in a full mating ritual before the female lays another single egg. But the male does not fertilise these later eggs, and the growing tadpole promptly eats them.

Opportunistic predators

Opportunistic predation is known in many species of quite diverse sorts of insects. The adults of some species of plant-eating grasshoppers and crickets are notorious carnivores whenever they get the chance. Often they are cannibals, and commonly scavenge on dead bodies of both vertebrates and invertebrates.

Another related and widespread, but usually overlooked behaviour of young arthropods (insects, crustaceans) is that of eating the shell of the egg they have just hatched from. However, a recent study of the caterpillars of a butterfly which feed on brassica plants in Brazil has shown how important a supplement this is for animals which otherwise eat nothing but plant material. And highlighting this is a twist to the behaviour that occurs if these flush-feeding caterpillars hatch on old, poor-quality leaves in the field, or in large batches of asynchronously hatching eggs in the laboratory. The first caterpillars to hatch do not stop at eating their own shell but quickly turn to cannibalising neighbouring eggs.

The shell of an insect's egg is pure animal protein, and in the case of these young caterpillars, amounts to a meal equal to 50 per cent of the neonate's weight. So the Brazilian scientists reared batches of these caterpillars on fresh leaves of kale without letting them first eat their eggshells, and contrasted their survival and growth with matched batches of caterpillars that were left to eat their eggshell before being given kale leaves. Many of those thus deprived did not survive beyond this first stage out of the egg, whereas most of those eating their eggshells did. What is more, the benefit of doing so flowed on. They weighed more, grew faster and became bigger butterflies which laid more eggs, than their less fortunate fellows. Clearly this initial meal of animal protein is of enormous value to these herbivores.

A related behaviour also widespread among arthropods, is that of eating their cast skins. All arthropods have to shed their external skeleton to grow, and they do this several times between hatching from the egg and metamorphosing to an adult. This, too, is a behaviour largely dismissed as being of any nutritional benefit to the animal. But a recent study in Oxford changed this perception. The work was done with colonies of the ubiquitous American cockroach maintained for 20 years at the University. This animal is not a herbivore, but a highly successful and adaptable omnivore – they will eat just about anything organic! Yet in nature they mostly have to get by on food that does not contain enough protein. To help overcome this shortage they have a large gut fauna of bacteria, and also carry specialised endosymbiotic bacteria that recycle the roaches' waste uric acid to usable amino acids.

The growing nymphs always eat their cast skin after each moult while only some of the newly emerged adult insects do so. Adult females, however, do so significantly more often than do the males.

The scientist who did this study found that the cast skins consisted of up to 87 per cent nitrogen that would be lost to the animals each time they moulted. By eating them, however, he found that they recycled between 60 and 70 per cent of this nitrogen. To see if the amount of protein in their diet influenced the frequency of this behaviour, he did two things. He raised one lot of cockroaches after killing their endosymbiont bacteria with antibiotics. Then he fed them, and a separate lot still possessing their endosymbionts, on a range of synthetic diets that differed only in the amount of protein they contained.

The results were fairly clear-cut. On the lowest nitrogen diet 90 per cent of the animals died before they became adults. All those that did survive, however, both male and female, ate all of their cast skins. By way of contrast none of the adults raised on a high nitrogen diet ate all of their cast skin. A third of them ate part of it, but 50 per cent of the females and more than 60 per cent of the males did not touch them. Those without endosymbionts ate significantly more skins than those not so deprived.

So, if you are a cockroach not getting enough protein in your food, or lack bacteria to help you gain more, you will attempt to counter this by recycling more of the otherwise wasted nitrogen in your cast skin. And if you are a female which has to produce young you will be even more inclined to do so, with or without the help of symbiotic bacteria. If, on the other hand, you have access to a nitrogen-rich diet, you will be much less likely to bother.

Small sap-sucking insects called thrips have been found to pierce and feed on mites' eggs whenever they find them on the surface of a leaf. Mostly they do this when they are fast-growing immatures. When they gain access to this supplement of animal protein they grow much more quickly, and far more of them survive to maturity. When adult female thrips are able to eat these mite eggs they live much longer and lay many more eggs than those confined to a pure diet of plant sap.

Cannibalism

As I have already alluded to several times, this propensity of plant-eaters to eat animal tissue can include eating one's own kind – cannibalism. A fine line can be drawn, of course, between being nourished by your mother's body (e.g. placenta, milk, trophic eggs, mucus, empty eggshells) on the one hand, and on the other eating your siblings, or (possibly your own) young. However, this latter form of more 'conventional' cannibalism is something that is, contrary to the belief of many, also widespread and common in nature. It is not a rare event seen only in animals under extreme stress such as over-crowding. It can be found in all forms of animal life, from single-celled microbes to mammals. And it takes many forms. Mate eats mate, parents eat their young, the young eat their parents, or more commonly their siblings.

Nor is it what many ecologists would have you believe, a so-called 'self-regulating' device to control the size of a population below the maximum size it could achieve by consuming all its available food. It is not a device to reduce numbers. On the contrary – it serves to increase the numbers that survive. It does this by using more of the food in the environment than would otherwise be used, and concentrating it into fewer, but successful animals. When food is short, without cannibalism some might survive; with cannibalism more will survive.

But it is not just any old food that is thus used more effectively. Ecologically cannibalism is just a special form of predation, functioning to increase access to animal food – to help alleviate the chronic shortage of protein in the natural world, and convert more of it to the next generation than could be the case without cannibalism. So it is not too surprising to find that cannibalism is most common among herbivores. Nor that it is largely restricted to breeding females and their fast-growing young. These are two points in the life cycle of any animal, herbivore or carnivore, when access to an enriched supply of protein is vital.

For example, it is fairly well known that a female praying mantis will eat her mate, often while he is still copulating with her. But she is much more draconian than this. Having been successfully impregnated, she continues to produce sex pheromone, so that more males are attracted to her. As fast as they arrive she promptly catches and eats them before they can mate with her! Research has established that this additional protein food increases both the number and viability of her eggs. And, incidentally, recycles to the next generation protein that would otherwise be lost in now-redundant males!

As we have seen termites are herbivorous, albeit with the aid of microbes. Yet their workers will kill and eat many of their own kind when nitrogen in their food is very low, and especially when the nest is producing the winged reproductives which must leave the nest to attempt to establish new colonies. And all cast skins, the injured and the dead in the nest are quickly devoured.

Many female mammals will eat their young (or resorb them as embryos) if food is short and they themselves are at risk of starving. Ecologically it is better to abandon the attempt to breed, recycle the young that were anyway doomed, and try again when conditions are favourable.

However, cannibalism by the young is even more ubiquitous. They may eat their mother entirely (some spiders do this) or part of her (as milk or egg-yolk for example). On the other hand they more commonly eat their siblings. Caterpillars reared together in small containers will eat their smaller fellows – and their larger ones once they start to spin a cocoon. At this stage they are unable to walk away or defend themselves, and their still-active brethren will quickly pounce on them. I have watched in horror as this happened in small cultures of hard-to-replace insects I had been trying to rear.

I related how garden snails which hatch first will eat unhatched eggs, and how evolution has refined this form of cannibalism in other species (both vertebrate and invertebrate) which produce unfertilised 'trophic' eggs, which are duly eaten by the young that hatch from the fertilised eggs. There is a more bizarre variant of this behaviour. Some sharks and salamanders have uteri in which their young grow. Here just a few of their embryos develop large jaws with teeth and proceed to eat their less well-endowed siblings! In cases like these it is possible to argue that this is just an extension of the mother's body nurturing her young. Nevertheless, if we define cannibalism as eating some or all of the body of another member of your own species, then all such nurturing, placental and milk feeding included, falls under this heading.

In general two broad cannibalistic 'strategies' can be recognised – the 'Lifeboat' strategy and the 'Grazer' strategy. The first strategy is fairly obvious. As the supply of food declines, or if it is scarce to begin with, the strongest

Figure 4.1 Most people have heard that a female praying mantis will eat her mate, even while in the act of mating. But that is only the half of it. After successfully mating she continues to produce sex pheromone that attracts more males which she will promptly eat too! Photo courtesy of LE Hurd.

members in a population eat the weaker. In this way the available food is concentrated to fewer individuals, but they each get enough to mature and produce a new generation. This is clearly better than all getting an equal, but inadequate share, with the result that possibly none survive to reproduce.

The codling moth provides a good example. Normally there are not enough developing seeds in an apple to support the growth of more than one caterpillar. So if a moth lays more than one egg in an apple the larger (and usually first-hatched) caterpillar eats its sibling.

The same story holds for the young of many raptor birds. Usually two or more eggs are laid each breeding season, but only in very good years can the parents catch enough prey to fledge more than one nestling. The oldest and largest will appropriate the lion's share of the food brought to the nest, and eventually kill and eat or eject from the nest its smaller and starving siblings.

An indication that this behaviour has been around for a long time is the recently described case of a cannibalistic species of *Bacillus*. When bacteria run short of nutrients they stop dividing and begin to form spores – a condition in which they can survive, often for many years, in a state of suspended animation. It has been found in one species that once an individual bacterium lacks adequate nutrients and enters the stage that would lead to sporulation, it releases a chemical killing factor into its surroundings. This chemical stops other nearby bacteria from forming spores and causes them instead to disintegrate, releasing their contents. These provide additional nutrients for the killer bacterium, enabling it to postpone sporulation and continue to grow and replicate.

The grazer strategy, on the other hand, is more subtle. Most commonly this strategy is seen where the young eat food that is not available to the adults – and then they are later eaten by the adults – often their own parents. Again, this concentrates otherwise inaccessible protein to fewer but successful individuals. For example, very small scorpions can catch animals that are too small for their mother to catch. She then eats most of them, thus 'grazing' food which would otherwise not have been available for her to convert to the next generation. Similarly, the free-swimming larvae of land crabs eat small planktonic animals in the ocean; food that is not accessible to the adults on land. Then when these larvae transform to miniature crabs and come ashore, many of them are eaten by the protein-hungry adults.

There are many more examples of cannibalism in nature, all of which achieve the same end – recycle and concentrate limited protein to fewer individuals so that more of them survive to pass their genes on to the next generation. But let us finish by taking a closer look at a more sophisticated form of the grazer strategy.

If you went walking in a European forest early on a spring morning you could come upon a line of marching soldiers. Follow, and you would see them fan out onto a battlefield and begin to fight with soldiers swarming on from the other side. The fighting is deadly – but silent, for these soldiers are all female workers of the European wood ant. And their foes are other workers of the same species from another nearby nest.

Such warfare starts when the nests have become active again after winter, and workers begin sallying forth in search of food. Battles rage all day, workers returning each morning to the same battlefield to resume hostilities. War may last for a month, with casualties commonly exceeding 75 000 workers from just one nest on just one battlefield. And a nest may be fighting battles with several different nests at once on different battlefields. So it is an expensive business in terms of the loss of food-gathering workers from a nest.

But this is not just war for the sake of it, nor to defend territorial limits. All slain enemies are taken back to the nest and fed to the growing young grubs there. Warfare topped by cannibalism!

Why would neighbours of the same species do such things to each other? Surely it is counterproductive behaviour? And 'unnatural'?

Not so. The key lies in the need for a nest to get enough protein to raise the next generation. Battles start when a nest's demand for protein is at a peak

Figure 4.2 Like most of its kind, this South Australian scorpion will cannibalise its small young. In this way it can significantly supplement its lean diet by indirectly 'grazing' the prey of the young which are too small for the adult to catch. Photo courtesy of Adam Lockett.

– when it is raising its annual sexual brood of winged males and females that will later depart to establish new nests. This happens in the spring when the usual prey of the ants (other, mostly plant-eating insects) are not abundant enough to meet this demand. If workers did not feed other workers to their young, their next generation would either fail, or be damagingly reduced in size. And all nests have the same problem. So warfare is inevitable!

However, as soon as other insects become numerous, peace is restored. In summer, when insect prey is abundant, the ants convert this plentiful food to large numbers of new workers. Next spring, when prey is again scarce, these now-aging workers constitute a live store of food to be fed – by way of these wars – to the next sexual generation.

But let us not overlook cannibalism among humans. It is generally thought by most people – including some biologists who should know better – to be an abhorrent and 'unnatural' aberration brought on only in times of extreme stress or deprivation, such as shipwreck or the famous example of the survivors of a plane crash in Chile. On the contrary, for early people living as hunter-gatherers it was a common practice (albeit disguised by various religious or cultural justifications). Modern archaeological investigations are producing mounting evidence (such as finding human myoglobin from heart and skeletal muscle absorbed inside pottery cooking containers, and in fossilised human faeces) of cannibalism having long been a common part of human activity. Indeed, it is not so long ago that this was so – perhaps it still occurs today! A study in 1974 of 'pay-back' warfare and cannibalism among small isolated groups of Papua New Guineans showed that this behaviour contributed 10 per cent of the protein to the diet of these people who were living where game was in chronically short supply. The only difference between this and the story of the ants is that 'surplus' young male humans were being recycled rather than sterile female worker ants!

So, a general picture emerges; vegetarians are not really vegetarians – at least not when they are growing youngsters.

Why, then, is this access to animal tissue (or, as we saw in the previous chapter, nitrogen-rich micro-organisms) apparently so vital and universal for herbivores? And if not for grown adults, at least for females generating and nurturing their young; and for neonates? Simply because plant food – a vegetarian diet – just does not provide sufficient protein for the rapid and exponential growth of a young animal's body. And this carnivory is still necessary even though, as we have seen in earlier chapters, all herbivores have evolved a wide variety of behavioural, anatomical and physiological adaptations to maximise their access to what digestible nitrogen there is in their plant food.

There is, however, one unequivocal exception to this apparent rule. That is the caterpillars of plant-eating moths and butterflies (and, possibly, the immature stages of some grasshoppers and locusts). All live on an exclusive diet of the leaves of plants. Many are known to be fierce, cannibalistic carnivores when given the opportunity. Anybody, like me, who has tried to rear such animals in captivity soon learns this. But they can be (and commonly are) raised on nothing but a diet of their plant host. The only possible animal protein they can get is the shell of the egg from which they hatch. Most hatching insects routinely do this, and as we saw earlier, may depend upon such behaviour to survive or breed. But even if denied this source of protein, these caterpillars can survive, grow and reproduce through repeated generations eating nothing but plant tissues.

How can they do this? There are two possible clues, both of which await careful investigation by insect physiologists. The first is the very high pH of the gut of these larvae – often pH 10 or more, as high as pH 12.5 in one species of termite; a very caustic brew! But such an alkaline environment is extremely efficient at extracting virtually every last trace of protein from the ingested plant tissues; much more so than the acidic gut secretions of other animals. This may increase the efficiency with which they can extract nitrogen from their food just sufficiently to tip the balance.

The second clue – and possibly linked to the first – is the presence in leaf proteins of 'Rubisco' (Ribulose biphosphate carboxylase/oxygenase). This is an enzyme found in cell chloroplasts, and chloroplasts originated as ancient micro-organisms that became permanently incorporated into the cells of early plants. Rubisco is a protein made of amino acids still much the same as those found in present day free-living microbes. Maybe the capacity to gain access to this animal-like protein in leaves, combined with the super-efficiency of extracting the last remnant of it from the cells of the leaves, frees these caterpillars from the need to be carnivores when they are very young?

Nevertheless, the list of vegetarians which are not truly so is diverse, and growing. Why has it taken so long to discover this universal nitrogen-hunger of plant-eaters? Because, I suspect, in Science, as in everyday life, so often we do not see something until we go looking for it. I am reminded of the (possibly apocryphal) story about the anthropologist investigating the diet and feeding habits of a Polynesian tribe in the South Pacific. He could not understand why the children were so fat and healthy. From his detailed recording of what they ate at meal times once they were finally weaned (usually not until they were three to five years old), he found that they ate practically nothing but very starchy vegetables: their elders ate what little meat was available! As a consequence they should have been suffering from quite severe protein malnutrition. However, what he had not observed was what the children ate between meals – insect grubs, shellfish, crabs and other small marine invertebrates, and seeds.

5

Feeding the favoured few

Territorial behaviour

Anybody who has spent time at the seashore will be familiar with limpets, those hard, flattish shells stuck fast on the rocks. Seeing one of these, motionless on its rock, you may not think it a particularly bright or aggressive sort of an animal. But in nature, when we look carefully, we are often surprised.

Limpets graze on a crust of algae growing on the surface of rocks, rasping it off as they move forward. Individual limpets maintain a specific area of the rock as their territory, and defend it against all comers – not just other limpets, but any of several other species of grazers. If a limpet encounters another grazer, it lowers the forward edge of its shell, and repeatedly strikes and shoves the intruder, until it either falls off, or is pushed outside the territory.

If a predatory snail – one that would attack and eat the limpet – shows up, however, its response is quite different. It raises the forward edge of its shell, and then brings it down sharply on the snail's soft foot; a behaviour called 'stomping'. Usually one such stomp is enough to make the carnivorous snail quickly retract its foot, let go, and fall off the rock. On the other hand, if our limpet encounters an inert object, it simply feeds around it. Remarkably varied and discriminatory behaviour for such an apparently 'simple' animal, don't you think?

Apart from seeing off potential predators, then, defending a territory sequesters a supply of good food from other grazers. The thicker and more luxuriously the algae grow, the smaller is each territory, and the more limpets there are in one place. The biggest limpets with the best territories produce the most offspring. Many smaller ones have to make do with eking out an existence in places where there is little food, or it is of poor quality. But if a good territory falls vacant, one of them will quickly take it over.

There are very many animals, from tiny invertebrates to large mammals, which, like these 'lowly' limpets, maintain a territory and show complex behaviour in defending it. Nearly always this serves to exclude others from a limited source of food. Some will argue, however, that there are many other reasons why animals might claim and defend a territory.

It is often said that a territory is maintained to defend a place to nest. Yet we find that at times when food is very abundant and many more females are breeding, such places are at a premium; there are not enough for all trying to nest. Nevertheless, every female will find somewhere to attempt to raise her young.

For example, one of the famous finches of the Galapagos Islands, the large cactus finch, was thought to nest exclusively in holes in cactus plants (hence its name). But during a time of very high and prolonged rainfall generated by the unusually strong 1982–83 El Niño, their insect prey became super-abundant on the lushly growing vegetation. In response they bred repeatedly, over a much longer period than usual, in places where they had never before been seen to breed, and fledged four times more young than usual. Holes in cactus plants were soon all taken, but this did not stop them. They made their nests just about anywhere; that year more than half their nests were in trees. What had seemed a limiting resource was not.

In Australia there are several species of native ducks that normally nest well above the ground in tree holes. When there is lots of food for them, however, and very many are breeding, there are not enough tree holes for all. Then those that miss out will nest on the ground where their eggs have little more than a screen of grass to protect them.

In the dry interior of Australia when there is lots of fresh green grass after good rains, feral rabbits breed continuously. Then all the subordinate females are driven from the security of warrens by the dominant animals. Nevertheless, with all that good food to eat, they continue to breed, but now have to drop their young in shallow burrows hastily dug in loose sand. With both the ducks and the rabbits their unprotected young are much more likely to be eaten by feral foxes. But the attempt is worth it if only some of this increased production survives.

Interestingly, titmice in the Netherlands illustrate the converse of this, and further emphasise the primacy of food over a place to nest as the key limiting resource. These small forest-dwelling birds normally nest in holes in trees, but in today's carefully managed Dutch plantations such holes are rare. So, routinely, artificial nest boxes are set out for them as it is believed that they protect the trees by eating caterpillars that can defoliate the trees. But often not all of these boxes are occupied. Furthermore, the number that are occupied varies according to the 'richness' of the habitat – more of them are used in mixed broadleaf plantations where there are more caterpillars than there are in pure pine plantations. And within these differing sites, more or less boxes are used from year to year as the number of insects varies with changing weather.

Another, more 'natural' example of the number of nesting sites that are used being dictated by the amount of food available is that of the kangaroo rats which live in the Chihuahuan Desert of Arizona. These animals construct extensive breeding mounds wherein they are protected from weather and predators, and where they store their principal food, seeds. Once built these mounds will last for decades, but in hard times many of them lie empty and neglected for years. After unpredictable good summer rains, however, there is a great increase in the amount of food available for the rats and their numbers rapidly rise. Then old abandoned mounds are quickly rehabilitated and reoccupied.

Very commonly territories are established by male animals. They do this, however, not to defend a supply of food, but to sequester one or more females from being mated by other males. First – or exclusive – access to the best females to raise his offspring ahead of those of his rivals is what matters to a male, thus passing his genes, not theirs, to the next generation.

For a female, on the other hand, there is always a surplus of males from which she can pick and choose. What matters much more to her, if she is to be successful in passing on her genes, is access to the best protein food in the environment. This she needs in her own diet to maximise her ability to nurture her foetus or egg, and for the nourishment of her young through their early exponential growth. And this is what most territorial behaviour is about.

Various hunting birds like eagles, hawks and owls provide clear and well-studied examples of this function of territoriality. The more prey there is for a pair to catch, the smaller their territory will be, and the more nestlings they will fledge each year. In poor, larger territories, or in years when there is little in the way of prey, they may not raise any young – in extreme cases not even establish a territory. As we saw briefly in the last chapter, in some species there is a further mechanism for ensuring that fluctuations in the amount of food available is used to maximise the number of young that can be raised to maturity; they hatch two or three eggs over several weeks. In good years they can feed all the nestlings, and all of them survive. But in tough times if they continued to try and feed all their young, none would get enough, and all would die. In this situation the oldest – and largest – nestling will kill and eat, or eject from the nest, its weaker siblings. The young of the Australian (laughing) kookaburra are notorious for this ruthless behaviour. But this way at least ensures that one 'favoured' chick may survive – better than none doing so.

A good example of the way in which territorial birds can respond to weather-driven changes in the supply of their food is that of the Andean condors of Peru. These large, long-lived birds feed, like African vultures, on the dead bodies of mammals. In the high Andes there is a fairly constant and adequate supply of this carrion, and the birds breed continuously. In the deserts of the coast and foothills, however, food is much more limited. On the coastal desert there are enough dead marine animals washed ashore for condors holding territories there to maintain a low level of breeding. In the foothills, on the other hand, where the average annual rainfall is only 9 cm, food is scarce. So scarce that in drought years the condors do not breed at all, even though they continue to defend their nesting sites and maintain their pair bonds.

But in El Niño years there are significant changes in the deserts. Rainfall increases dramatically (during the 1982–83 El Niño more than 4 *metres* fell in the foothills in nine months). This causes a great increase in the number of deaths among the large herds of free-ranging livestock that are run in the foothills. Pairs of condors which have not bred for years quickly initiate breeding in response to the resulting bonanza of carrion. On the coast, however, it is a different story. Changes in ocean currents generated by El Niño drastically reduce the number of marine carcases washed ashore and the condors holding territories there stop breeding.

There is a final twist to this story. Because of the vagaries of the El Niño cycle, desert-dwelling condors can seldom produce enough young to maintain their numbers over the long term. They can only maintain their populations because they are replenished by immigration of some of the 'surplus' offspring produced by the ones that breed continuously in the more consistent climate of the highlands.

You are probably familiar with many examples of birds forming territories. If you live in Australia you may well have a pair of magpies that raise their young in a territory near your home. And you might have noticed the annual aerial dogfights that precede the carving up of territories before the breeding season starts. But have you also noticed the roaming band of birds that have missed out on a territory? Males and females both, they do not breed but remain ever alert and ready to seize any vacancy should an occupant die. And next season they will be fighting along with the rest to try and gain both a territory and a mate. Usually, however, there are a few dominant birds which hold on to the same territories year after year, driving out their fully grown young before they start breeding again. I have a female magpie that comes to my back door and feeds from my hand. Each year she brings her mate and, in due course her young, to learn to share in the handouts. For 16 years now she has nested in one or other of the tall trees near our house. In that time she has lost at least three mates but has still managed to raise at least one young each year. Among all the magpies that come to my door, she feeds first; all others defer to her. Last spring, after 17 years, she failed to return. But one of her daughters – with a handsome new mate – is now the matriarch coming to my backdoor.

On the other hand you may not be familiar with some of the less apparent animals which defend territories, including many which live entirely on plant food. Let me tell you about some of them.

In America there are tiny aphids, only 0.6 mm long, which form galls on poplar leaves when these are expanding in the spring. Two aphids will fight long and hard with each other to decide which one will settle closest to the base of the leaf. They push and shove, sparring like boxers, sometimes for days, until one (usually the smaller) either falls off or moves away. Aphids which manage to settle at the base of the biggest leaves claim the best territories – areas no more than 3 mm long – where they develop the biggest galls within which they grow to maturity and produce many more offspring than their less fortunate sisters. This is because this position on the leaf is where they can insert their sucking mouthparts into plant sap of the best quality – sap richest in soluble amino acids, the essential ingredients for successful production and growth of their young – as it is imported into the growing leaf.

But what of the losers? Some find smaller leaves, or a position further out along the leaf from the prime site on the best leaves. There they may produce a few young. Most, however, exist for a while as 'floaters', searching for sites to settle, or challenging holders of territories, but soon dying. They are the 'doomed surplus' that miss out on the chance to survive and breed.

Another form of territorial behaviour by insects is perhaps more nearly akin to the siblicide of nestling raptors. Codling moths, or at least their caterpillars, are all too familiar to anybody trying to grow apples. The moths lay their eggs in the just-formed new fruit and the tiny caterpillars feed on the developing seeds in the ovary. But unless the apple has many large seeds they provide enough food for only one caterpillar; more than one sharing and all would die. Then, the oldest – and largest – one eats its siblings. This too, has been mentioned in the previous chapter.

Figure 5.1 A mere 0.6 mm long, these newly hatched aphids fight each other, often to the death, for possession of the prime site at the base of a poplar leaf to form their galls. Photo courtesy of TG Whitham.

Still in the domestic garden, there is another, but more sophisticated example of insect territoriality. There is a very tiny wasp that lays its eggs in the eggs of the green vegetable bug – that large foul-smelling beast that attacks your tomatoes and beans. Again more than one wasp grub per egg and none would survive, so the first and biggest one eats any others. But this rarely happens because each female wasp, after she has laid one egg in the egg of the vegetable bug, wipes her ovipositor back and forth across the top of that egg, depositing a pheromone on its surface. This signals to any other female that the egg is already taken and they respond by not attempting to lay their own eggs in it.

Another fascinating invertebrate example is that of a species of ground-dwelling spider which lives in desert grasslands in Arizona and New Mexico. The females are very territorial. Each builds a funnel in which to shelter, and a sheet of non-sticky webbing in front of the funnel. Any insects moving onto this sheet are instantly detected by the spider in her funnel and she quickly jumps out and grabs it. But beyond this web each spider maintains an area which she fiercely defends against all other females of her species. This includes not just holders of neighbouring territories, but any of the many 'floaters' – individuals that have not been able to gain a territory for themselves. These animals hide wherever they can in cracks and crevices, and subsist by trying to steal prey from the webs of holders of territories. If a territory becomes vacant a floater will quickly move in and take it over, but most do not survive for long, let alone produce offspring.

In places where insects are sparse there are many fewer spiders and they defend territories that are much bigger than those of spiders in the best places. What is more, spiders holding the smallest territories where there is most prey produce 13 times more offspring than their sisters on the poorest land.

An experimenter put some of these spiders in enclosures and either fed them daily with insects or deprived them of food for four weeks. The starved ones increased the size of their web. Well-fed ones, on the other hand, not only reduced the size of their webs, most let them disintegrate altogether. They had quickly learned that an abundant supply of good food arrived every day, so just came to the mouth of their funnel at feeding time to take the insects offered to them! (Who said welfare dependence is confined to humans?)

In contrast to this sort of deliberate experimentation, every so often human manipulation of the environment unintentionally reveals what is happening in nature. One such case involved a study over the past 24 years of serow in northern Japan. Serow are goat-like ungulates that live in mountainous areas of mixed broadleaf forest and plantations of conifers where they browse on the leaves and twigs of a variety of shrubs and bushes. In this

Figure 5.2 The size and number of territories of the Japanese serow expand and contract markedly in response to changing availability of their food plants generated by the clear felling and regrowth of the conifer plantation where they live. Photo courtesy of Keiji Ochiai.

stable habitat males and females maintain separate solitary territories, and vigorously drive off any other individuals of their own sex. They have no predators, and hunting was banned long ago. Over the years of this study, some adults died or vanished, and all young kids left their mother's territory as soon as they were mature. Yet the number and size of occupied territories remained nearly constant as, consequently, did the density of the serow population. One very severe winter slightly reduced their numbers, but they quickly recovered. Then, in one part of the study area, some plantations of mature conifers were clear-felled. The response was dramatic. Many shrubs grew on the cleared land so that there was as much as 150 per cent more food for serow there than in undisturbed parts of the study site. And very soon there were up to six-fold as many serow, occupying territories that were three times smaller, on the cleared sites. However, as the newly planted conifers grew and started to suppress the shrubs on these sites, the number of serow began to decline. But it is estimated that it will be 20 years before numbers fall to those present before the clear-felling.

What, then, is happening in all these examples? Is territorial behaviour, as many ecologists would have us believe, a form of population control which reduces the numbers in a population so as to conserve limited resources for the use of future generations? Or is it making sure that what resources are available are shared out equally among those seeking to use it? Neither explanation is correct. Altruism is a purely human concept. Territoriality has nothing to do with conserving resources for the next generation nor of sharing them among the present one. To do either of these things would, in any case, be evolutionary suicide. Others intent on using every bit of food they can get would quickly outstrip those altruistically 'trying' to restrain themselves so as to ensure others got a fair share, or, equally improbably, conserving some for future generations. And in doing so those 'selfish' individuals, by using all the food they can, would leave more offspring than the altruists, ensuring that their genes would quickly dominate the gene pool.

In a world where there is rarely enough food – and more particularly, enough protein food – for all those trying to eat it, one way to make sure that what food there is gets used as effectively as possible, is to channel the limited supply to only some of the many individuals seeking to use it. All but a 'favoured few' are denied the use of a scarce resource for the production of the next generation. And only these few breed successfully. Some of the rest may find an inferior site and produce some offspring, but most – the 'doomed surplus' – may manage to subsist for a time, but eventually die without reproducing. But, of course, they are not a surplus at all. They are a reserve. Whenever the amount of food available increases, some of them will establish new (and usually smaller) territories to make use of it. And should a vacancy arise in an established territory one of them will quickly take it up. In either case the available food is put to good use – producing more of their own kind.

In all the examples I have told you about, and in hundreds more, this is what territorial behaviour achieves. And, though it may seem paradoxical to some, in so doing it maximises, not reduces, the number of individuals that the environment can support in each generation.

Finally, interesting illustrations of the prime place of protein food in the maintenance of territoriality are some responses to 'unexpected' concentrations of good food – both man-made and natural – in the habitat. Black bears in North America are solitary, mostly herbivorous, and strictly territorial. They will not entertain another bear near them. But where they can gain access to humans' rubbish dumps – a rich source of protein food – they suspend this intolerance of their own kind while all gather to partake of the feast. A similar thing happens with other, equally solitary bears. Grizzly bears gather to feed on the annual runs of salmon up the rivers, and even entirely carnivorous polar bears come together and 'socialise' at places where there are a great many seals to catch. In all cases otherwise aggressively individualistic animals 'happily' tolerate others nearby while all are busy harvesting a bonanza of food in a world usually distressingly short of food. But, in each case, all is not entirely sweetness and light. Within a group they quickly establish a hierarchy where physical dominance decides who gets the lion's share of the pickings.

Nor is this behaviour confined to bears. In northern Norway stoats – also fierce holders of individual territories – congregate at rubbish dumps of tourist lodges. What is more, they persist at these places when in the surrounding countryside their numbers have crashed following the collapse of populations of voles – their natural food. A similar story emerged of wolves congregating and feeding at night at rubbish dumps in several European countries. Also in Europe red and roe deer are frequently fed throughout the winter to help maintain their number for hunting. Where these animals congregate at the feeding stations established for them, wild lynx soon learn to congregate for an easy meal. Lynx, wolves and wolverines behave in the same way, and inflict heavy losses on reindeer which are herded together and fed through the winter in northern Scandinavia.

There is another story that I must tell here. It is a further example of how wild animals subsisting on a less than adequate diet quickly learn to congregate to take advantage of food concentrated in their environment by human activity. But it is also an illustration of the unintended flow-on consequences that this may generate. It is not the sort of story that would ever get into the scientific literature, but it was told to me by a respected colleague who assures me that it is true.

Wild feral goats have become so abundant in the Flinders Ranges of South Australia that they constitute a serious threat to native plants and marsupials. One means of reducing their numbers, and at the same time making some money from them, has been to round them up, halal-slaughter them, and ship them frozen to Middle East countries. There it eventuated that some very wealthy families cook and present a whole goat at each meal. Any of the carcase not eaten at that meal is not considered worthy of being kept and is thrown onto a dump in the desert. There troops of local baboons gather and feast upon this unexpected and persisting supplement to their normally uncertain scavenged diet. This large extra input of animal protein into the baboons' diet has resulted in a huge increase in their numbers; so much so that they have become a serious pest. An example of an attempt at environmental conservation in one place causing an environmental problem in another!

Social dominance hierarchies

We have seen something of the way in which territorial behaviour operates to sequester the limited amount of good protein food in the environment to only a few females and their growing progeny. And how this meant that all others, including previous young, must be expelled from the territory.

Well, the same function is fulfilled by increasingly complex social structures where such expulsion does not happen, or is delayed, or does so only now and then. Here, animals of two or more generations live permanently in a group and 'share' the resources in their habitat. At its most complex this specialised behaviour produces the social groupings we see in the mammals – and ultimately our own societies. But there is no equal sharing in these societies. A strict hierarchy of dominance by a few individuals prevails, and these dominant few get the best food; and only they breed. The concept of democracy and equal sharing and opportunity for all is a recent human cultural overlay – and then all too often honoured more in theory than practice. It certainly plays no part in nature.

But first let me tell you a story about the saving of an endangered species of bird, the Seychelles warbler, reduced to the point where it was found only on a single island, Cousin. These are very strongly territorial birds. They are also what are called 'cooperative breeders'; that is, the young remain in their parents' territory when they are fully grown and help feed the next generation of nestlings. However, they do not themselves breed.

The size of the warblers' territories and the success of their breeding depend on the supply of insect food available in a territory. In a high quality territory with much food young birds stay often for several years. Most of these only leave to breed when a vacancy appears in another good quality territory. Any young bird which waits like this for a good home produces more offspring in its lifetime than any which leave earlier and have to breed in a mediocre territory. Any bird which leaves to attempt to breed in the poorest territories has little chance of ever producing any young.

Destruction of the warblers' habitat had reduced them to only 26 birds on one island. However, careful management to restore the habitat produced a spectacular recovery. In five years the entire island was again covered with territories. Not until that happened, however, were any young birds seen to stay in their parents' territory and 'help'. Prior to that all had immediately left to establish their own territories and breed. Even then, it was only in a few of the richest territories that any young birds were staying.

But after 14 years the population had ceased to grow and stabilised at about 300 birds. By then 'helping' was widely observed in territories all over the island, and from then on the only young birds to leave home were those that could find a vacant territory in which to start breeding.

Once the population was no longer growing, researchers transferred some young birds to two unoccupied islands. These, too, had been renovated to again provide suitable habitats where the birds could establish viable territories. On these vacant islands the warblers bred much sooner, more quickly and more often than those birds left on the original, and now fully populated island. But as had happened there in the early years, none of their young stayed home as 'helpers'; all left immediately to establish new territories. Until, that is, all the high quality territories were occupied. From then on young birds born in these good territories began to stay as helpers. And they did this even although there was still abundant unoccupied space apparently available for new territories. However, investigation revealed that the unoccupied areas were of very low quality, containing little insect food.

There are three points to this story. First, it illustrates that the availability of habitats which contain adequate food dictates the number and size of the territories these birds can establish, and the way their young behaved in these territories. This would never have become apparent had the islands never been denuded, and remained fully occupied by warblers. Second, it revealed that the young birds stayed at home only so long as it was to their advantage to do so, not out of any altruism towards their younger siblings or their parents. Finally, it illustrates the link between strict territorial behaviour on one hand, and hierarchical social groups on the other. In the former, once all good habitat is occupied by territories, all young are driven out – usually to soon die, or at best hang on at the periphery in the hope of a vacancy cropping up. In the latter, young are allowed to stay. This increases their chances of eventually moving into a vacant territory where they can breed successfully. And this is of evolutionary advantage to both them and their parents – all get to pass on more of their genes than they would if the young were kicked out straightaway.

But in all such animal societies the young that remain and reach maturity in the group are themselves prevented from breeding, and are in every way kept subservient to the dominant, breeding adults and their latest young. They get less and poorer quality food, less safe places in the habitat, and are first to die when things get tough. Also they may 'earn their keep', as it were, by making themselves useful to the dominant ones. As I have just discussed, they may help to feed and care for the young. Or they may groom their betters, guard the group against predators and conspecific enemies – and be the first to die from attacks by either. It is better to tolerate such second-class citizenship in the hope of one day getting the chance to pass on your genes by usurping a previously dominant but aging or sick individual, or finding a new place where you can be king pin. The alternative of immediate departure means almost certain death, and no chance to pass on your genes.

In America there are small ground-dwelling herbivorous rodents, called marmots, which live in systems of burrows. Their story illustrates a further step along this spectrum, or continuum of complexity of social behaviour. And it illustrates once more that a lack of good food in the environment is the pivotal factor driving these associations.

Their burrows are aggregated on better quality sites: those with most food. Each set of burrows is home to a varying number of marmots, which defend it against other groups of marmots. The better the quality of the site the more marmots in a set. All the animals living in a set are females, and in each there is a single dominant one. She alone breeds. All the other marmots in her set are either her sisters or her grown daughters. Such a group is called a matriline. Young males leave home before they are sexually mature, and adult males maintain a separate hierarchical social structure among themselves. Each dominant male defends one or more matrilines against other males, and breeds with the dominant female in each.

But even in the best of worlds, continued increase in numbers means that sooner or later somebody has to go. Then it is the newly maturing young and those of lowest rank in the group (the two are usually synonymous) which are driven out. Most will soon die; they are, again, the so-called 'doomed surplus'.

This is well seen in populations of feral rabbits in Australia. Because of the immense environmental and economic damage they do their ecology and behaviour have been extensively and intensively studied. They live in warrens – complexes of burrows dug deep beneath rocky outcrops or tree roots, where they are safe from predators and the vicissitudes of the weather. Around each warren the occupants mark and defend a territory. Each warren contains a group of three to five females, one of which dominates the others and claims the best place in the warren to nest and the best food in the territory. The warren also houses two to three males, one of which is dominant. He mates with all the females. The others rarely get a look in, apart from a few lucky couplings – and then only with the lowest ranking females. So both sexes maintain a strict social hierarchy wherein the dominant animals gain first access to the best food and do 90 per cent of the breeding.

All this can change dramatically, however. From time to time widespread rains turn this hot dry land into a sea of green growth. Then all rabbits – even those lowest in the hierarchy – breed continuously and prolifically. Warrens are renovated and enlarged and new ones are established, but still there are more and more females breeding, and no room left for all of them to nest in a warren. This, however, does not stop the breeding. Young females of low status that are kicked out of the warrens dig quite shallow burrows away from the warrens in which to give birth. As we saw before, these are easily dug up by foxes, and few of their young survive. Yet, in spite of this, so long as the green grass lasts, very many more young survive and mature and the number of rabbits continues to increase.

If we look still further along the continuum of social structure we find increasing complexity of that structure.

Within the group the actual behavioural interactions and their complexity will differ. In some, like a herd of deer, it is relatively simple. A single male, having driven off all competing males, will sequester a varying number of females and mate with all of them. At its most complex are the social groupings of primates where a group consists of several related families.

A socially structured group may live in a geographically defined and defended territory, or it may range more or less widely over an area which is frequented by other groups of the same species. Each group will defend itself against conspecifics from all other groups while maintaining a strict and ruthless hierarchy of dominance and unequal distribution of resources within its own group. In every case the same basic function is fulfilled. Dominant males mate with all, or all but the lowest ranking females. Dominant females get first pick of the best males to mate with and commandeer the best food. The available resources in the environment are efficiently concentrated for the wellbeing of a favoured few. While this may seem cruel in human terms, in evolutionary terms it is good, for it means that those individuals that are best at appropriating and using the limited food pass on their genes at the expense of those which are not so good at it. It is also good ecologically, for it sees to it that as many individuals as possible – more than would have been possible in the absence of a social structure – are produced in each generation.

Some experiments done at the University of Wisconsin more than 50 years ago, and long since forgotten by most, clearly showed that the amount of food available is what drives territorial and social behaviours – and limits the growth of populations. Small populations of wild-caught house mice were established in a series of large rooms in an old warehouse. The rooms were lined with sheets of galvanised iron 60 mm up the walls, higher than any mouse can jump, and turned in along the floor to prevent them gnawing out beneath it. Each population was provided with many large cardboard boxes as nesting sites along with an abundance of cotton waste and shredded paper as nesting material. Each room had a permanent and unlimited supply of drinking water. All populations were fed on a diet known to support vigorous growth and breeding of mice; a rich mix of grains, powdered protein rat pellets and powdered dried meat.

Two separate experiments were run, and each lasted for a year. In the first one the mice were given a fixed and limited amount of food each day, in the second they were provided with a constant surplus of food.

In the first experiment the numbers of mice increased rapidly, but the population abruptly stopped growing when the 'food crisis', as the experimenters called it, was reached; when the daily amount of food was no longer sufficient to allow all mice to eat as much as they needed. From then on their numbers started a slow, steady decline. The survival of young had been very high until the food crisis. At this point, however, all but one of the few young born soon after died, and no female showed any evidence of pregnancy from then on; there was a complete cessation of breeding. All adults continued to obtain enough food to survive and to maintain their body weight, but all became physiologically and behaviourally sexually inactive.

Space was not limiting. More than half the animals in the room were crowded into just one 'house' while many of the other boxes remained empty and with unused nesting material. But the young could not disperse out of the room, which is what happened in unconfined populations that were given the same daily ration of the same food. And the females in these unconfined populations continued to produce young. So the key was the amount of food per animal.

Experiments with laboratory rats and mice have shown that quite moderate fasting will stop their oestrus cycle without them losing any condition. An experiment with American white-footed mice demonstrated the same thing – but with one important difference. Individual females were fed equal ad lib amounts of food, identical except that some received diets containing less protein. Just a 10 per cent drop from an intake of protein on which 100 per cent of females bred, stopped nearly all breeding within three weeks. Yet none of the mice – even those with 30 per cent less protein in their diet – had lost any weight, nor was the level of their carcass fat reduced. So an individual mouse's capacity to breed is immediately and drastically reduced when not just the amount of food, but the amount of protein she eats, is reduced by so small an amount, or for so short a time, that it has no detectable influence on her physical condition.

In the second Wisconsin experiment there was a very different outcome. Numbers increased steadily and at times exponentially, but the population again stopped growing quite suddenly. This time, however, it was not because of any decline in the number of litters being produced, but because practically none of the young survived. By the time this happened crowding had reached the stage where there was general social instability. There was much intense aggressive activity with fighting very prevalent. Fighting was mostly, but not entirely, between males, 75 per cent of which were mite-infested, scabby and with open wounds. Females had ceased to defend their nests which were constantly being encroached and damaged by other adults. Worse, they were eating their pups. This cannibalism was the major cause of the death of the young – and of the population ceasing to grow.

But matters didn't stop there. A gradual decline in numbers followed. And this was due to continued social discord among the crowded animals. All semblance of the normal male hierarchy was gone so that mating became disrupted and chaotic. Many males pursued and fought over any female in oestrus, and few managed to copulate properly. At and around feeding sites fighting and harassment became so bad that the amount of food being eaten per individual fell below the level known to be necessary for females to keep breeding. This, and the diminished rate of conception as many failed to be impregnated in the first place, led to a decline in the numbers pregnant and the rate of births. Throughout all this the rate of mortality of mature mice remained unchanged, but that of newly weaned sub-adults increased markedly as they missed out in the scrabble with their elders for the food.

So, whether a population does not have enough food or enough space, if the young cannot disperse it will cease to grow.

In the first situation, left to itself the population would have continued to be stable as long as the food supply did not change. Over time there would be minor changes in the number of animals, as once sufficient adults had died to allow the survivors to get enough to eat, some breeding would replace them. But, of course, in nature there is never a steady and unchanging supply of good food. Quite the reverse. Both the amount and quality of food available for natural populations is constantly changing, often by a great deal.

The second situation is an aberration because it does not arise in nature. There is never a permanent excess of high quality food available, and young will always disperse, even if to their death. Forcing this doubly impossible situation on the animals resulted in their behaviour and physiology becoming totally inappropriate. Evolved to maximise the use of what food there is in the habitat to produce as many young as possible, and to see to it that they then dispersed to find new sources of food, they were now pathological. Presumably a population left in this situation would have eventually adjusted to such conditions, but I imagine it would be a long process and even crueller than the experiment proved to be.

# 6

Inefficient killers

Most people believe that predators are efficient and all-powerful killing machines that keep the numbers of their prey well below levels that would be possible without their depredations. But most people are wrong! As is often the case in nature, what seems superficially to be so is not when we look more carefully. Predators are generally inefficient.

The food of carnivores, unlike that of herbivores, is a concentrated source of protein. It exists in discrete 'packages', the bodies of other animals. If there are enough of these available to be caught there is enough good food to support reproduction and growth. Unfortunately, nearly all these packages are very mobile, and thinly and patchily spread in a variable environment. They may not be all that easy to find. Furthermore, even when they are quite plentiful and visible, prey animals are equipped with a range of sophisticated behaviours for avoiding becoming the victim of a predator. So, what seem superficially to be lots of easily caught animals are actually mostly difficult or impossible to catch.

All this means that predators are confronted with an environment every bit as harsh and inadequate as are herbivores. They are chronically short of food. As is the case with herbivores, however, this shortage does not normally impinge on adult animals. Most carnivores in their prime can catch enough prey to sustain themselves. The problem arises with females trying to reproduce, and with their young when they are undergoing their rapid early growth. At these times a lack of protein becomes critical. So much so that females commonly fail to breed, and most young that are produced soon starve. Precious few of their kind ever reach their prime.

You don't believe me? Let me give you some examples.

I have watched tiny, freshly-hatched New Zealand katipo spiders trying to eat aphids that I fed to them, being repelled and distressed by the aphids' caustic secretions sticking to their mouthparts and legs. I have seen how frequently an insect hitting an orbweb spider's web breaks free before the spider can get to it. On the other hand I have seen spider-hunting wasps caught by the very spiders they attacked. I have watched small parasitic wasps, which must sting and stun their prey before laying an egg in it, flung away by the potential victim's wild thrashing, or repelled by its secretions or vomit. Large caterpillars and adult aphids are virtually immune from attack because of these tactics as these wasps can successfully do battle with only the very small early stages of their prey. But even then they are far from efficient in their behaviour. When a female wasp encounters potential prey she will frequently go into a frenzy, thrusting repeatedly but aimlessly in all directions with her sting, more often missing than hitting her intended victims. Observing larger parasitic wasps that seek out newly pupated caterpillars in leaf litter on the forest floor, I found them to be trebly handicapped. First they search randomly with their feelers among the litter, and frequently pass within a hair's breadth of a potential victim without detecting it. Then, when they do find one, they are commonly flung away by its thrashings and never locate it again. Finally, even when they do manage to stay with one long enough to try to sting it, their sting cannot penetrate the thick tough integument of the chrysalis. To be successful, a wasp must find a chrysalis that has just transformed from the caterpillar and when its skin is still soft enough for the sting to penetrate.

These examples, however, are only some of my own casual and unquantified observations – 'mere anecdotal evidence' my critics would call them. So I will now relate evidence that I have taken from properly conducted and published studies.

Lions and other inefficient killers

Everybody knows about the mighty African lions: king of the beasts, renowned supreme hunters of anything from small antelopes to huge cape buffalo. Well, for a start, a male lion rarely hunts. He leaves this to his pride of lionesses. He is much more interested in fighting other males for the right of access to lionesses!

The lionesses are the providers. They cooperate to ambush, run down, overpower and kill their prey. But they are, nonetheless, not all that good at it. There is usually no absolute shortage of potential prey animals in their habitat. On the contrary, there are enough to feed very large numbers of lions year round, but because of their inability to catch most of them they suffer a chronic relative shortage of food. With larger animals they can mostly catch only the very young. In practice this means that only in the wet season when large herds of grazing animals are concentrated together, and producing young at the same time, is there a brief flush of relatively easily caught newborn calves. For the rest of the time they must try to isolate and bring down an adult beast. But unless that animal is old, or injured, or momentarily inattentive, their chance of a successful kill is small. Most of what to us looks like an abundance of prey is simply not accessible to them. Therefore lions eat whatever they can catch, and kill the easiest prey. This means that for much of their lives they are forced to hunt only small vulnerable species – things like rodents. These comprise the bulk of their diet. Yet even when they concentrate on hunting these easiest-caught prey, more than half of their attempts at capture fail; and their average rate of success, overall, is only around one in three to seven attempts.

Consequently their abundance is strictly limited by the amount of catchable prey in their habitat. Most of the time they are chronically hungry, and the lack of food impinges most on the young. The cubs, burdened with poor nutrition from the milk of their underfed mothers, and later unable to get a fair share of the pride's infrequent kills, are the first to succumb to starvation. Few survive to 12 months.

There is another sort of lion most people do not know about: ant-lions. These are insects – the young of certain lacewing flies. They get their name because they mostly catch and eat ants, although they will devour any small insect that comes their way. They are what we call 'sit and wait' predators. Each newly hatched ant-lion digs a pit – looking like a little volcanic crater – in dry sandy soil. You can often find them on dusty paths or similar dry places in summer. The ant-lion buries itself at the bottom of its pit with just the tips of its long, hollow, poison-laden mandibles above the surface. There it stays, perfectly still, until an incautious ant blunders over the edge of the pit. Then it immediately leaps into action, throwing sprays of sand up the side of the pit, creating little avalanches that will bring the ant tumbling down to within reach of those jaws through which the ant-lion quickly sucks the contents from its body.

But, as a careful study in Sierra Leone showed, there are two things mitigating against any one young ant-lion getting enough food to survive and grow. First, even where ants and other ground-dwelling insects are quite abundant, the chances of any of them actually falling into a pit is not large. It may be a very long time between meals – too long and you die of starvation. This is the common lot of most recently hatched ant-lions: most never get to eat at all. Second, the tiny ant-lion cannot handle every animal that does fall into its pit. Any that are longer than 2 mm are too big to subdue; shorter than 1 mm, and they are too small to be grasped in its jaws.

For the few ant-lions that do survive and grow bigger, the increased size of their pits and the speed with which they can consume their prey improves their chances of growing to maturity. Paradoxically, however, the very largest, near-mature ones, become almost as vulnerable to a relative shortage of food as they had been as new hatchlings. Most of the animals falling into their pits are now too small for them to grasp and eat, and very few of the larger ones, which they can process economically, ever fall into their pits. So, once again, they are at severe risk of starving.

But to return to the cat family. The staple food of the Canadian lynx is the snowshoe hare which is renowned for its huge fluctuations in abundance. When there are lots of hares there are lots of lynx. They thrive on the abundant supply of easily caught young hares. But when the hares are scarce the lynx are forced to eat mice, voles, squirrels and a miscellany of other small prey – even insects. Yet this shift in diet is not sufficient to stop a severe fall in the total amount of food they can get; their body weight falls drastically and most are unable to breed. Those that do, have smaller litters and hardly any kittens survive from spring to the beginning of winter. Of the survivors 95 per cent die early, when they are still dependent on their mothers for food, from malnutrition and starvation.

Another cat familiar to us all is the lion's smallest cousin, the domestic cat. It is now established worldwide, and in some unlikely places, as a feral wild animal. As such it is similarly limited by the amount of food for breeding females and their young. No matter what sort of habitat it occupies, anywhere in the world, it has the same reproductive potential to rapidly increase in response to increased availability of food. However, because the level of attainable food is very different in different localities, it persists at very different densities.

By and large they are not terribly good at catching enough prey. But where catchable food is abundant, cats quickly increase. For example, they were recorded as being as dense as 200 animals per square kilometre in the docklands of Portsmouth, in England, where human refuse provided a superabundance of food. On the other hand, on subantarctic Campbell Island, where most birds have been eliminated and there is no alternative prey, there are too few of them for meaningful densities to be gauged. In between these extremes a wide range of densities has been recorded. In a Californian waterfowl refuge they reached nearly eight per square kilometre. Again in the subantarctic, on Macquarie Island, there are as many as eight or nine per square kilometre, because there are plenty of easy-to-catch young feral rabbits and ground-nesting petrels to eat. When they were first introduced to the equally bleak Marion Island, in the Indian Ocean, where there are lots of these burrow-inhabiting mutton-birds, they were recorded at 16 per square kilometre and increasing at an annual rate of 25 per cent.

Feral cats in forests and National Parks in Australia and New Zealand are mostly pretty thinly spread; at best two or three per square kilometre, frequently fewer than one. They subsist on anything they can catch – mostly feral rodents, but often little other than a few insects. Most of each year's crop of kittens soon vanish: dead from starvation.

In Australian semi-arid grazing country their staple food is the introduced rabbit. Here a study to measure the extent to which feral cats and foxes can regulate the abundance of feral rabbits finished by demonstrating the ultimate dependence of both predator and prey on the supply of their food.

Persistent and concentrated shooting of cats and foxes over large tracts of land allowed numbers of rabbits to increase. In matched control areas where they were not shot these predators kept the rabbits' numbers down – but only so long as the rabbits' breeding was depressed by a shortage of food. Following good rains, and the consequent flush of new grass over these huge areas, rabbits everywhere exploded in abundance. Their predators in turn bred much more successfully, but were nevertheless quite unable to contain the rapid increase in numbers of rabbits. Until, in the subsequent drought, the grass all died off and millions of rabbits began to starve. Then the many young cats and foxes thrived on the easily caught starving rabbits. But it was a short-lived bonanza. Soon all but a few rabbits were gone, most of the cats and foxes starved, and their numbers plummeted.

This same red fox which is feral in Australia is similarly limited by its ability to catch enough of its staple food in its native Europe. In the northern Scandinavian areas this is largely voles, and its numbers rise and fall with the numbers of voles.

An experiment with the Arctic fox in Swedish Lapland showed these predators, too, to be limited by their food and their capacity to catch it. Like those of the red fox, their numbers, and more significantly, the number of surviving cubs they produce each year, go up and down with the numbers of voles in the area. Researchers placed fresh meat near selected dens to supplement the prey these foxes could catch, and compared them with foxes in other dens left to fend for themselves. The foxes getting the supplement produced many more cubs at weaning than those in the control dens. But this was still not enough. The scientists tagged 65 cubs during this study but only three of these survived to the following spring. Without a continuing assured supply of food after they were weaned, few of these still-growing cubs would have escaped starvation.

Another canid, the wolf, in North America preys mostly on moose or various species of deer. They usually can kill only the very young, but packs will run down old and debilitated adults. When the ungulates produce lots of young many more wolf cubs survive, and this is the major factor affecting changes in the size of their populations. But a good deal of the time adult wolves must rely on catching small mammals – even insects – to subsist. Then few of their cubs survive. Anybody who saw David Attenborough's TV program on wolves saw a good illustration of both how inefficient wolves are at catching even small prey and the resulting slow starvation of their cubs that ensues when they can't catch enough.

As we saw with the ant-lions, close observation of invertebrate predators shows them to be every bit as fearsome in action as any cat or wolf – but just as inefficient!

Bungling invertebrates

If you have ever been out and about in a field of tall grass on a frosty morning you may have been lucky enough to see hundreds of small orb webs strung between every available grass stem, and with droplets of dew condensed on them shining in the rising sun. While no doubt entranced by this very beautiful sight, you probably did not notice what was happening in the grim world of nature. Normally these webs are as good as invisible. When highlighted like

Figure 6.1 The orbweb spider's web is not as efficient a catcher of insect prey as is usually thought. Only a small fraction of the insects flying in the air will chance to fly into it, and many that do are so large that they fly right through it. Of those that are trapped, quite a few will struggle free before the spider can grasp them. Photo courtesy of David H Wise.

this, however, you can see that they create a virtual blanket cover of all the spaces between the grass plants. Any air moving between those plants, and any insects flying in that air, are almost certain to encounter one or other of those webs. A pretty efficient way to catch most of the insects passing through that field it would seem. But for any one young spider only a tiny fraction of those insects will enter its particular web; possibly none if the insects are thinly spread. And then only some of those that actually hit its web will stick. Many are so big and heavy that they go straight through. Only small ones are trapped. Moreover, quite a few of those will struggle free before the spider can get to them. There is much potential food, but little of it is accessible to any one individual. So, for much of their short lives young orbweb spiders are at best hungry, at worst starving. Few of the myriad of them in that field would have survived. They died because their tiny fragile webs could not catch enough small insects often enough to sustain them. It has been shown experimentally that some can subsist between rare catches by eating pollen caught in their webs – an interesting reversal of the herbivores' ploy of eating animal protein to tide them over a lack of nitrogen in the plant.

Jumping spiders fare no better. Either sitting in wait in a flower, or actively hunting, they never catch more than 40 per cent of the insects they try to catch, and frequently record a catch rate of less than 2 per cent. A study of a species of jumping spider in Maryland, USA showed very clearly how inefficient they are. These spiders wait in flowers and attack all insects which visit them, making up to 20 attempts an hour. Their chief prey (in terms of food gained per capture) is bumblebees. The next most important are a small species of fly. The spiders were recorded to catch only 1.6 per cent of the bumblebees they attacked, and 39 per cent of the flies – about one fly a day and one bumblebee every three days: they are missing most of their potential food, and having minimal influence on the numbers of their prey in the process. Worse, however, the scientist studying them found that they were being far less efficient than they could be. If, instead of attacking all comers, they just concentrated on bumblebees, they would have gained over 7 per cent more food for their efforts. And if they further confined their hunting to times when the prey were most accessible they would have got 20 per cent more food. But to expect such sharp thinking from a spider is asking a bit much. They are programmed to catch what they can of whatever they encounter, not to optimise their hunting. As a result they take little of the food present in their environment, and mostly go hungry.

The same story emerges for another predator that most people see in their gardens, the praying mantids. These insects are also sit-and-wait hunters, standing motionless until a potential meal comes within sight. Then they slowly turn their heads in an uncannily intelligent-looking way, watching closely until the victim comes within reach of their fearsomely spined forelegs. Then they strike with lightning speed, hold the catch firmly, and start chewing away while it continues to struggle.

Yet life is not as simple as it seems. Few mantids get to be big enough for us to witness such an event. Although surrounded by a great abundance of insects, most starve as tiny beasts just hatched from the egg case. There are two reasons: first, because few potential prey come within their reach, and second, because they fail to catch most of those that do, because either their strike misses, or those that do come close enough are too big and powerful for them to handle. Field experiments in America with the introduced Chinese mantid have demonstrated that, while living in fields which contain an abundance of prey, over 95 per cent of young mantids are dead within two weeks of hatching, and fewer than 1 per cent survive to maturity. In laboratory experiments newly hatched nymphs denied food were dead in less than five days. If, however, like the young spiders, they were fed nothing but pollen, they could survive much longer – almost as long, in fact, as young mantids fed a surfeit of small vinegar flies. Although they took three times as long as the vinegar fly-eating control group to grow, and only achieved a quarter of their weight, nevertheless, access to pollen in the field meant some could survive between rare meals of insects.

As a result of all this inefficiency, the density of predators in any habitat is, like that of herbivores, dictated by the amount of food they can wrest from their environment – often only a fraction of what is there. And the corollary is, of course, that they cannot regulate the numbers of their prey. It is the other way around.

Staying in the world of the invertebrates – populations of the sheep tick in the Scottish Highlands illustrate the point well. Ticks must have three meals of blood to complete their life cycle. Not necessarily sheep's blood; any bird or mammal – from human to tiny field mouse – will do. But there is a catch. Female ticks lay their eggs in the dense mat of dead grass on the ground. When the young ticks hatch they must climb to the top of a blade of grass and wait for a warm-blooded creature to pass close enough for them to grasp it and feed on its blood. Then, after each meal, they must drop off their host and return to the sward of damp grass to moult to the next stage before again climbing up and waiting to latch on to another passing meal. While they wait they dehydrate, so if they are not soon successful they must return to the humid sward to resorb water. And then climb back and try again. After several such unsuccessful journeys they die of starvation. Even in a paddock with many sheep, and numerous small mammals and birds, the smallest of which could supply a meal for very many ticks, most ticks die before a prey comes close enough to grasp. Of all the ticks which hatch in a field, less than six per cent of them will get three meals. The rest starve because of a relative shortage of food in the midst of an absolute abundance.

A similar story emerged from studies of the African tsetse fly, the carrier of the deadly sleeping sickness. The larvae of these flies develop to maturity within the bodies of their mothers, and so depend entirely for their nutrition on the diet of the females. These females feed on the blood of wild and domestic mammals, and must take many meals to nourish a series of larvae to maturity. Like the Scottish ticks, once they have found a host it has sufficient blood to feed thousands of them, but they must find a new animal each time they need a meal. Between times the flies must seek out the relative safety and shade of the bush while they digest their meal of blood. They have the advantage that a tick does not of being able to actively fly and hunt for their prey, yet animals may be so widely scattered and hard to find that many may die before they can find a meal, and many will be lucky to get enough feeds to mature their young.

All this was dramatically demonstrated by an experiment to try and control tsetse fly in one part of Tanzania. Hunters were employed to shoot most of the game animals over some 1500 square kilometres. After five years thousands had been shot, but there were still enough animals left to provide many times more than enough blood for all the flies that had been there at the start of the experiment. However, they were now so scarce, and so sparsely spread, that the chance of any one female fly finding a meal was reduced to near nil. So much so, in fact, that the experiment was successful; it drove the tsetse flies in this locality to extinction.

Ground beetles are another group of fierce hunting invertebrate carnivores. These fairly large, beautiful, shining black or green creatures run actively around searching for other insects to eat. Once they find one they quickly overpower it with their large mandibles and eat it. But they are not all that good at finding prey in the first place, even when there are enough insects nearby to feed many beetles. Extensive studies of several species of these beetles on the northern heathlands of the Netherlands have shown why. Like all invertebrates, these beetles do not have the complex nervous systems necessary to develop the sort of hunting skills that vertebrate animals have. Essentially they are programmed to move about at random until they encounter a victim. Having found and devoured one they are then stimulated to hunt actively in their immediate surroundings. If another is not soon found, however, their behaviour quickly decays back to random walking. This is an effective strategy for finding some food when it is patchily distributed in a heterogeneous environment – but not for finding more than a small fraction of it. Many beetles, especially when they are very small ones just hatched from an egg, starve to death before they ever find a meal. Few will survive to maturity. They suffer a relative shortage of food in the midst of a great abundance of food: they simply are not efficient enough at finding it.

Food supply is the key

Finally, back with vertebrate animals, the story of the Australian pelican is another good illustration of how populations of predators are limited, just like herbivores, by a shortage of food. They, too, have the in-built capacity to explode in numbers immediately food becomes plentiful: but just as quickly die off again when the supply disappears.

Most of the time pelicans live on the coast, breeding only occasionally if the fishing is good enough. However, when the dry salt-flat that is Lake Eyre South, 650 km inland, fills with water flowing from heavy rains in Queensland (a fairly rare and unpredictable event), pelicans quickly migrate from as far afield as the coasts of South Australia and Papua New Guinea, to feed and breed on the enormous number of small fish teeming in the lake. Their numbers explode in response to this sudden superabundance of good food. During one such event in 1989–90 there were estimated to be 200 000 birds there – about 80 per cent of Australia's total population of pelicans! But before long evaporation caused the lake to rapidly dry out, becoming eight times as salty as the sea and killing the fish. Then thousands of unfledged chicks died, and adult birds deserted nests with unhatched eggs, and departed

Figure 6.2 Pelicans breed on Australia's Lake Eyre on the rare occasions when it is filled with water and teeming with fish. They rapidly increase to outbreak numbers, but once the water starts to dry up, and the fish die, the pelicans too die in their thousands. Photo courtesy of Tony Lewis.

for the coast. Any fledged young accompanied them, but any bird unable to fly was doomed to stay and starve. But in this short time of superabundant food a great many new pelicans had been produced.

The pelicans' story is repeated for their cousins around the world. The brown pelican of the Californian Gulf eats practically nothing but anchovies. In a good year when these are reasonably abundant, adult birds can get enough to maintain themselves. At such times there are enough anchovies in the Gulf to feed all the young that they could possibly produce. However, these fish are very patchily distributed. So the adults often cannot find plentiful shoals within foraging range of their breeding colonies. Then their young are malnourished or starved simply because their parents are unable to gather enough anchovies, fast enough, to fuel their rapid growth.

But in El Niño years real disaster strikes. The warm current flowing towards the American west coast suppresses the upwelling of the cold nutrient-rich waters. This sudden blocking of the flow of nutrients passes rapidly up the food chain, via plankton, invertebrates and fish; ending in widespread starvation for many species of seabirds. For the brown pelicans the disappearance of the anchovies means that many nestlings starve, fewer are fledged and many nests are deserted. In especially poor years whole nesting colonies abandon their nests and even adult birds starve.

Far away in the Rift Valley of Africa small populations of the great white pelican persist wherever there is permanent water containing enough fish to permit some breeding. However, infrequently and unpredictably, heavy rains will fill vast areas of what are usually dry salt pans in the valley with fresh water. Then there is a huge bloom of algae and invertebrates in the water. In turn the fish breed in millions. Just as happens in Lake Eyre, so do the pelicans, feeding on the fish. Then, just as happens there, once the water evaporates the fish die, and so do the pelicans. First the young as the adults abandon the whole colony, leaving the chicks to starve, but then adults, too, as the supply of fish continues to shrink.

This same species of pelican also breeds on Arel Island in Mauritania. There it undergoes the same boom and bust fluctuation in its fortunes, but annually, and driven by ocean currents, not rainfall. Each year in July there is an inflow of warm water from the Gulf of Guinea and this brings hordes of pelicans that feed on the abundant small fish and crustaceans in this water. They lay eggs in successive waves through until the end of November. But early each December an upwelling of cold water displaces the warm water to the south and with it their prey living near the surface. The cold water is almost devoid of surface-dwelling fish, so there is an abrupt end to the food supply for the breeding pelicans. Adults and any young that can fly quickly depart, leaving deserted nests and thousands of unfledged chicks that soon starve.

So we see that three species of pelican, in widely separated parts of the world, are responding in the same way as the rabbits do to a sudden and great increase of their food: and they collapse in just as spectacular a manner when the supply of food dries up. Not only do these stories (and the others in this chapter) clearly illustrate that populations of carnivores – just like the herbivores – are limited by their food, they make it quite clear that carnivores do not regulate the abundance of their food. Furthermore, in the case of the pelicans (and big cats and wolves), because, unlike rabbits, they do not have any predators attacking them, there is nothing to invoke as a 'regulator' of their abundance.

There is one exception to all this. That is where humans have introduced predators to places where they never had been, and where they encounter prey which have never evolved ways of avoiding being caught and eaten by a predator. In these sorts of situations – and there are many of them around the world – the introduced predator has the capacity to eliminate some species – and has often done so.

One good example of this is the plight of the Australian malleefowl. This bird has evolved a very special way of raising its young. The adult birds build large mounds of dead leaves that generate heat as they decay. Once this process is under way the hens lay their eggs in the mound. By carefully monitoring the temperature and adjusting the depth of leaves in the mound, the adult birds can keep the temperature constant and at a level for successful incubation of the eggs. Once the chicks hatch they dig their way out of the mound and run off into the bush where they must fend for themselves. Introduced European foxes have become adept at waiting by a mound and catching these young birds when they first surface. They have little chance of escape and studies have shown that, where foxes are sufficiently abundant, they will repeatedly kill all young malleefowl emerging from a mound, season after season. Even for these long-lived birds, extinction is assured should this situation persist.

One attempt in NSW to prevent this slaughter consisted of repeatedly spreading poison baits for the foxes. But this proved to have little effect. While very many foxes were killed, there remained a hard core of 'bait-shy' individuals. They would never take a bait, and maintained sufficient vigil alongside mounds to continue to clean up most of the young chicks. It was only when it was realised that the foxes could maintain this pressure because they were not dependent on the young birds for food that success was achieved. Rabbits are the natural prey of the fox, and provide its staple diet. So as long as there were plenty of rabbits the malleefowl chicks were simply more easily caught 'icing on the cake' for the foxes. Once the considerable local population of rabbits had been removed, and vigorous baiting of the foxes was maintained, their numbers were reduced sufficiently to allow a significant numbers of chicks to survive.

Across the Tasman Sea a similar situation exists. European stoats were long ago introduced to New Zealand (along with ferrets and weasels!) in the mistaken belief that they would control the burgeoning populations of introduced rabbits. They have since spread and colonised native forests on both the North and South Island where they are a serious threat to the continued existence of several native birds, especially flightless ones like the kiwi, attacking their newly hatched young. There are no rabbits in these forests, but the stoats' staple diet is made up of two other introduced animals, the house mouse and ship rat, both also everywhere adapted to life in New Zealand's forests. So, once more, an introduced predator has a base diet to sustain it while it puts constant pressure on a much more easily caught alternative prey – young, defenceless native birds. In some forests recruitment of juvenile kiwis stays in continual decline until the numbers of stoats can be reduced by 80 per cent. Such reduction is virtually unsustainable in the exceedingly rough country where these forests grow. And it is impossible when the occasional outbreaks of rats and mice allow the stoats to breed up to large numbers. Then, when the outbreaks collapse, many now-starving stoats put even greater pressure on the kiwis. Careful experimental manipulations of the numbers of stoats in these forests have also revealed that the stoats have little influence on the changes in abundance of the rats and mice. This is generated by changes in abundance of the food of the rodents; southern beech seed masts and the associated increased numbers of herbivorous insects.

7

Plagues, outbreaks and the tyranny of weather

Previously I have mentioned that the weather has had a decisive influence on the abundance of animals. Ultimately, this must always be the case. In the last analysis, weather is the arbiter of the fate of all organisms on this earth. It dictates the conditions that hold sway in all habitats on earth, and how and to what extent those conditions change over time. Thus the weather determines what sorts of plants and animals, and how many of them, can live in each and every habitat.

At the simplistic level this is obvious, and could be thought to be a trite statement. Only animals and plants that can tolerate cold temperatures can survive in the Arctic. Only those adapted to live with little access to water can live in a desert. And we observe that organisms have evolved a myriad of specialised adaptations to enable them to cope with different levels of heat or cold, wetness or dryness, in their habitats.

But the influence of weather on the changing abundance of organisms is both more subtle and yet more straightforward than this direct impingement. While its influence on plants is usually direct, its influence on animals is often not. More usually it is indirect, via its influence on plants, with the response of the plants then influencing the success of animals that eat them. In turn, the success and abundance of animals that eat the herbivores will depend upon the success of the herbivores.

Weather's dramatic effects

A long-term study of several species of Darwin's finches on the Galapagos Islands provided a particularly compelling – and very dramatic – demonstration of how this link of weather–plant–herbivore–carnivore operates.

These finches breed during the short, hot, wet season, although often, on these relatively arid islands, there is little rain and no birds will breed at all. Some species eat mostly insects, some, mostly young soft seeds. All eat insects in the breeding season and feed them to their young. They have no predators.

The study was conducted on some of the smaller islands (Daphne, for example, is only 40 hectares) so that every one of the birds present was identified and counted, and all their nests were found and observed. The investigation had been running for about 10 years when the extreme 1982–83 El Niño hit the islands.

The wet season started much earlier and finished much later than usual, with *10 times* more rainfall than the previously recorded *maximum*. Many more plants grew than in most years, and they were larger and more lush, and flowered and fruited continuously. They produced 11 times more seed than the previous season, and the proportion of soft new seeds (the preferred food of the finches) rose from 25 to 80 per cent.

There was an equally dramatic rise in the abundance of insects living on the plants, especially of the caterpillars the finches preferentially feed to their young. These were six times as abundant as usual, and present for four times longer.

The finches responded predictably to this explosion of their plant and animal food. They bred continuously and for twice as long as the usual four-month breeding season. They produced four times as many clutches, laid five times as many eggs and fledged four times as many young as in a normal year.

Figure 7.1 During very high El Niño-generated rainfall on the Galapagos Islands plants grow more luxuriantly, set much more seed, and very many caterpillars grow on them. Galapagos finches, confronted with this sudden super-abundance of food, breed for longer, lay many more eggs and fledge many more young than in a normal year. They quickly increase to 'outbreak' numbers but just as quickly die off once the rain ceases and the food supply dries up. Photo courtesy of Sonia Kleindorfer.

Many more females than usual bred, new territories were established in previously unoccupied areas and nests were built in places where they would not normally be built. Some species bred on islands where they had not previously been recorded doing so. Young birds, which normally would not breed until they were two years old, bred before the end of the season.

However, because of the incessant stormy weather with heavy rain and strong winds, many nests were deserted and large numbers of nestlings died. Not withstanding this great increase in mortality, by the end of the season numbers of finches were exceptionally high and most of them were young birds – a veritable outbreak of finches!

At the same time numbers of other birds, in particular the Galapagos mockingbird and cuckoo, also increased markedly. But, one man's meat …! While the birds' numbers increased so dramatically, more than 60 per cent of their near neighbours, the Galapagos marine iguanids, starved to death. Rising El Niño sea temperatures around the archipelago had killed most of their algal food growing on the rocks. And it was two years before any of the surviving females bred again.

Meantime the party was soon over for the finches. Dry years followed the El Niño deluge and high mortality of both adult and juvenile birds soon reduced the population to pre-El Niño proportions.

This dramatic explosion and collapse in numbers – this outbreak – gives a strong clue as to why there are not nearly so many finches in 'normal' times. They have the capacity to increase, but not the wherewithal to do so – they have access to much less food. There are many similar examples, with more being described all the time. And, increasingly, the role of major changes in the weather like the only recently understood El Niño-Southern Oscillation (ENSO), are being recognised as the driving force behind these fluctuations in abundance of animals, and how this flush of food travels up the food chains, each level dependent upon changes in the one below.

Such is now the case with outbreaks of a dozen or so species of small rodents in the drier parts of South America. There are records of these outbreaks dating back over 450 years to the middle of the 16th century from Chile, Argentina, Brazil and Peru. All of these, it is now emerging, have been generated by El Niño events producing peaks of exceptional rains where usually there is little rain. And with each peak there is a great flush of growth and seeding of grasses, especially several species of bamboo, and ephemeral herbs – staple food for small rodents. In semi-arid northern Chile rains associated with the 1991–92 El Niño produced a three-fold increase in these seeds and herbs, and some 11 species of rats and mice which eat them exploded to more than 20 times their usual levels of abundance. The major predators of these rodents – hawks, owls and foxes – showed a delayed response to this flush of their food, more than doubling in frequency 12 months later.

In many parts of inland Australia there is usually little rain. So for much of the time this land is a dry, arid place, where the only plants that can survive are deep-rooted and drought-resistant perennials. Nevertheless, a great many species of animals still manage to live in this habitat. The most obvious of these are several species of large kangaroos. Usually they persist at very low levels of abundance, spread thinly over great tracts of land. A few adults will find sufficient food to survive, so that the population does not go entirely extinct within the greater habitat. But they do not breed, except for a chance few that are lucky enough to find a rare place, such as a natural waterhole (and since European settlement artificially created ones), where there is sufficient green feed growing to allow some young to be produced. At the same time the kangaroos' predator, the dingo, ekes out an equally pitiful existence. Small numbers of them persist by scavenging on the odd carcase and catching any prey they can – usually only insects; now and then a small rodent. Nor do they breed – or if they occasionally do, their pups have slim hopes of surviving long beyond their birth.

But now and again a great deal of rain can fall in quite a short time. Then the desert blooms. All manner of plants, evolved to remain dormant for long periods without rain, sprout and grow rapidly, producing enormous quantities of green growth, flowers and seeds. And kangaroos eating this lush, nutritious food start to breed. They are exquisitely adapted to respond very quickly to any such sudden flush of good food: and to keep on breeding so long as it lasts. Females during dry times carry embryos in their uteri, but these do not develop. As soon as there is green feed for the females to eat, however, the embryos develop, are born within days, and climb into the pouch where they start suckling. Immediately after birth each mother is again impregnated so that as soon as the young joey in her pouch is big enough to leave, the next young one is born.

And so it goes. So long as there is green feed kangaroos will keep on breeding, and their numbers will rapidly explode. And, of course, the dingoes, suddenly presented with a great supply of young, easily captured kangaroos to eat, produce numerous and large litters of pups nearly all of which will quickly mature and start breeding.

Inevitably, however, this scenario cannot last because there will be no more rain; maybe not for years. So the plants set seed and die. Then hundreds – thousands – of kangaroos, most of them not yet fully mature, starve and die as their food disappears. For a short time this further boosts the supply of easily caught prey for dingoes. But soon, too, their turn comes. Life returns to its usual state. Once more few kangaroos – or dingoes – can live in this harsh environment.

Many ecologists, in spite of the evidence of these binge and bust events, still believe that predation by dingoes, rather than the amount of food, is what is restricting the numbers of kangaroos in arid Australia. They often quote as evidence a study which showed there were many kangaroos on the side of the famous dingo fence where the dogs are shot out, but few if any on the other side where they roam unmolested. A recent re-evaluation at the same study site, however, revealed a different story. It so happened that here the fence coincided with a natural boundary. On one side is a large basin into which streams, flowing after rain, drain and terminate. On the other there is dry, sand dune country which does not receive any of this runoff. Green herbage grows in abundance in the basin, and lasts for some time after rains, and kangaroos gather there and breed. Further south of the basin there is one stream which sometimes, when there is a lot of rain, flows through the fence into the dingo country. There it creates a small flood-out; and that is the only place on that side of the fence where green grass grows. Over four years of observation this was the only place where kangaroos were found on the dingo side of the fence.

Reinforcing the conclusion that it is the availability of green feed, not the absence of dingoes, which decides where kangaroos will gather and breed, there is an area some 600 km further away from this site where the landscape is well watered on both sides of the fence. Here there is no difference in the numbers of kangaroos on either side.

In this self-same environment the introduced European rabbit and its predators, the feral red fox and domestic cat, have become established. And,

Figure 7.2 The availability of green feed, dictated by infrequent rainfall, not attacks by dingoes (A) is what limits the abundance of the red kangaroo (B) in arid inland Australia. Photos courtesy of Peter Bird (A) and Rob Drummond (B).

like the kangaroos and dingoes, the numbers of rabbits and their introduced predators fluctuate just as violently in response to weather-driven changes in the amount of green feed. So, too, incidentally, do the numbers of the native wedge-tailed eagles, which have found rabbits to be an ideal and easily caught food.

Another example from the dry centre of Australia is that of locusts. As happens in many parts of the world, with many different species of locusts, the Australian plague locust now and then appears in huge numbers out of the dry interior and swarms across pasture and crop alike, destroying all before it. During the usual times of drought in the interior a few small populations of locusts manage to find enough plants on which they can survive and breed. But mostly these plants, although still green, have finished growing and are less than ideal food for growing young locust hoppers: 90 per cent of them are dead soon after hatching. Once more, however, infrequent and unpredictable good rains generate vast areas of new green growth in the dry interior. Then hoppers hatching in this actively growing grass fare much better: only about 60 per cent of them die. This is, however, a huge, fourfold increase in their survival. When this happens over thousands of square kilometres, and for two or more generations of locusts, it is enough to generate a massive explosion in numbers which soon start the march out into farmland. And they are nearly impossible to stop, even by killing untold millions of them with insecticide. Today, knowing and understanding how and where these plagues are generated, we can monitor the weather and, when the rains first start, move to quickly kill the relatively few early hoppers until the habitat again dries up.

Two recently published examples arising from long-term ecological studies, while not finding such sudden or dramatic changes in numbers as the previous examples, nevertheless again reveal how the weather-driven supply of the new growth of plants determines how abundant an animal will be.

The first is a 40-year study of a population of wildebeest on the tropical African Serengeti. This is a very large, but essentially closed population (i.e. they do not migrate in or out of it) and the animals are grass-feeders. Most animals die before they are a year old, and the greatest mortality occurs in newborn calves. There are five large carnivores which prey upon wildebeest, lions and hyenas being the principal ones. Nevertheless their combined efforts play only a minor role in limiting the wildebeest population. It is food supply that is the primary cause of mortality of wildebeests – 75 per cent of deaths are due to malnutrition. And variations in the supply of food – green grass – is directly caused by variations from year to year of the rainfall. Parallel long-term studies of the wildebeests' two major predators, lions and hyenas, have revealed that changes in *their* numbers are determined, via changes in the survival of their young, by these weather-driven changes in the availability of their prey.

The second study revealed the same story for very different animals in a very different part of the world – birds of the grouse family which live in the cold of northern Europe. Their populations have been extensively and intensively studied in several countries over many years. Probably the best known of these is the red grouse, managed for centuries as a sporting bird in Scotland. Although grouse chicks eat insects in the first few weeks of life, they are otherwise strict herbivores, feeding on flush new growth of their food plants (in the case of red grouse this is exclusively heather). These birds go through cycles of abundance, and the conclusion from these studies is that the main cause of changes in their breeding success – and hence their numbers – is the variation in the recruitment of young birds into the population from year to year. And the level of this recruitment is due primarily to the quality and quantity of the diet of the hens and their chicks. This, in turn, varies in response to variations in the weather influencing the flush of new growth of heather and the abundance of insects.

All these examples are fairly straightforward cases of animals responding to an increase in plants driven by an increase in the amount of rain. But there are other, and often less direct ways in which changes in the weather drive changes in the abundance of animals by changing not just the amount of their food, but its quality. One such is the case of the Australian zebra finch. As I described in Chapter 2, these small birds rely on a supply of newly ripening grass seeds to be able to breed and raise their nestlings. They live in mobile populations which move over large home ranges in the arid interior, and are opportunistic breeders. At any time of year when there is sufficient rain for grasses to germinate and grow they will start to breed, and the heavier and more prolonged the rain, the more intensely and longer they will breed. Two months after rain – a month or so longer in the winter when the grasses grow more slowly – they start to nest. The first clutches of eggs hatch just as the first ripening seeds become available. Nestlings are fed exclusively on these. If there are follow-up rains they will continue with surges of breeding activity, each new hatching corresponding precisely with the onset of a new flush of ripening seeds. And they may continue to do so for long periods (in a particularly wet spell in Alice Springs they were recorded breeding continuously for nearly 11 months). On the other hand they will not breed at all over long periods when there is no rain, or too little rain.

Not surprisingly weather-driven changes in the availability of protein food extends beyond herbivores to carnivores, and their success can just as surely depend on these changes.

Some of the infrequent but exceptional rains in the dry Australian interior cause floodwaters to flow into Australia's inland Lake Eyre which is just a dry salt pan for most of the time. When the flooding is great, the lake is filled with nutrient-rich water. In this water bacteria and invertebrates eating them proliferate, as do the fish feeding on the invertebrates. Then, as related in Chapter 6, Australian pelicans flock from far and wide to eat the fish and breed in enormous numbers. Soon, however, the water evaporates and becomes highly saline. This kills the fish and most of the newly bred young birds starve.

Exactly the same sequence of events happens with the great white pelican on the large lakes of the great Rift Valley of Africa. There, too, infrequent but great rains cause flooding of huge areas of usually dry lake beds. However, as happens in Lake Eyre, the water evaporates, the food supply dies and so do the pelicans that have briefly enjoyed a period of rapid expansion.

Then think of the example of the Peruvian condors in Chapter 5. On the coast they ceased to breed when their food decreased, and in the desert their food supply boosted their breeding when it increased; both responses generated at the same time by the same El Niño affecting the amount of food for them to eat – but in different ways.

And back to pelicans again. The Californian brown pelican lives and breeds in many places along the American west coast feeding on the abundant schools of anchovies that live and breed in the nutrient-rich cold water upwelling from the deep ocean. But in times of El Niño this all changes. Warm water flowing from the west stops the cold upwelling, so phyto- and zoo-plankton don't grow, the fish don't breed and thousands of pelicans starve to death.

Ironically the reverse happens for the sub-population of the great white African pelican which breeds on Arel Island off the coast of Mauritania, and it is an annual event. Yet, as I related in Chapter 6, the outcome is the same. Cold water, devoid of small surface fish, wells up each year and drives out the warm water which is rich in small fish. The pelicans' breeding collapses, thousands of eggs are abandoned and immature chicks are left to starve.

Successful reproductive strategies

Not all carnivores are so profligate and 'wasteful' of their young, however. Some have, like the kangaroos, evolved the capacity to vary the number and timing of the production of young according to the prospect of success in an uncertainly variable environment. A particularly good example is seen with the European stoat. Stoats are small but ferocious predators which evolved in the northern hemisphere where their principal prey are voles, animals notorious for the wild fluctuations in their abundance.

But it has been intensive studies of them in the fiordlands of the southwest of the South Island of New Zealand which have revealed not only how dependent they are on weather-driven changes in their food, but how exquisitely they have adapted to the uncertainty of its supply. As shown in Chapter 6, stoats were introduced to New Zealand and are now a significant threat to the survival of several ground-dwelling species of native birds. But their staple prey, and key food resource, are feral house mice which have invaded all corners of the country. In the fiordlands they are usually fairly scarce, but each time there is a mast year of the native beech trees, and they produce huge crops of seeds, the number of mice explodes. And so do the numbers of stoats. There is a very close correlation between the number of mice in the spring and the number of juvenile stoats in the following summer. Numbers of stoats can vary from none in a year when mice are few to more than 12 independent young per female in years when they are numerous. And how many young stoats survive to independence depends upon the fertility of the females being adjusted to environmental conditions by a process of sequential juvenile mortality.

Stoats have evolved a very special and unusual reproductive cycle. Geared originally to their spasmodically variable food supply, the Northern Hemisphere voles, it proves to be equally as effective in the face of similarly fluctuating numbers of feral mice in the New Zealand bush. Young are born in the spring and females are mated while still in the nest and only five weeks old. But implantation of their fertilised eggs is delayed for 9 to 10 months. The numbers of these that eventually do or do not implant, the number of embryos which subsequently survive to full term or are resorbed, and the number of nestlings which survive to independence, depend upon the nutrition of the female and how many prey she can feed to her nestlings. If mice are scarce she will be undernourished, able to carry and mature fewer embryos, produce insufficient milk for her nestlings, and unable to catch enough food to rear them past weaning. In a very lean year none will survive.

Back in Australia, a small omnivorous native rodent, the dusky rat, lives only on the flat treeless floodplains of tidal rivers in the monsoonal north of Australia. These plains are inundated in the wet seasons by the monsoon rains. The extent of the flooding each year and how long the plains take to dry out afterwards all depends upon the amount and timing of the rain each wet season, and this varies greatly and unpredictably. And the abundance of the rats varies dramatically in response. At the end of the wet season the water recedes, and the clay soils of the plains start to dry out. Only then can the rats invade the plains and start to breed. The vigorous growth of plants on these drying soils provides an abundance of the rats' food of sedge corms, growing grasses, seeds and insects. The soils crack as they dry, providing a cooler microclimate where the rats can escape from the very high temperatures. How long they are able to breed depends upon the length of time the soil stays moist, and their food lasts. In some wet seasons with little rain, or very wet ones when the plains remain covered in water late into the dry reason, they may hardly breed at all. In a good year, however, when the soil remains moist well into the dry season and food remains abundant for longer, the rats maintain excellent body condition and will breed continually for as long as nine months. And when this happens their numbers explode, producing enormous populations in a very short time. The very high fecundity of these small animals – some have likened their capacity to multiply to that of insects – sees to this. Females start to breed when they are five to seven weeks old, and can produce a litter of four to 12 pups every three weeks. At that rate, if all young survive – and most of them will while there is plenty of good food – one pair of rats would have 400 descendants in just 26 weeks.

The chief predator of these rats is a water python. Like all predators these snakes concentrate in areas where their prey is most abundant; in this case where the soil stays moist longest and there is most food for the rats. In years when rats are abundant the pythons feed frequently, put on large fat reserves, and the following year most females produce young. Conversely, in years when rats are rare, most snakes are thin, and few females reproduce next year. However, given the rate at which the rats can breed when there is abundant food, it is no surprise that these pythons – and other predators which attack the rats – have little effect on the growth of their numbers at such times. The

Figure 7.3 Numbers of the Australian dusky rat (A) fluctuate as its supply of food changes at the dictate of unpredictable variations of monsoonal rains. In bad years they hardly breed at all, in very good years their numbers quickly explode to enormous populations. The rats' major predator is a water python (B). When there are few rats these snakes are thin and few females produce young. But when rats are abundant they grow fat and breed prolifically. Yet they cannot match the insect-like speed of their prey's breeding, so have little effect on their numbers. Photos courtesy of Thomas Madsen (A) and Peter Harlow (B).

predators can merely harvest a little of their own sudden food bonanza before it again shrinks to low levels.

Another large jump, this time geographically from the hot tropics to Macquarie Island and the Great Southern Ocean; from a short-lived 210 g rat to the long-lived 800 to 3700 kg Southern elephant seal; and from the rats' massive capacity to breed to females which can produce but one pup a year. Here we find another way in which weather is influencing the abundance of an animal through the supply of food for breeding females and neonates. These seals spend only 20 per cent of their life on land, coming ashore only to breed and moult. For the rest of the time they roam over thousands of kilometres and to great depths in the Southern Ocean hunting fish and squid.

By the end of the 19th century hunting had drastically reduced the numbers of these animals. From then on, however, hunting largely stopped, and by the 1950s they were again enormously abundant on Macquarie Island and other breeding grounds around the Southern Ocean. But then, inexplicably, in less than 40 years their numbers fell by 50 per cent.

Intensive studies on Macquarie Island are attempting to discover what might be causing this still continuing decline. The proximate cause was soon

Figure 7.4 If this Southern elephant seal's females feed well then their pups will be born fat and strong, and they will have abundant milk for them. Most will thrive. When ENSO-driven changes in the ocean cause a shortage of food for the females, their pups are born thin and weak, and many will not survive their first foraging trip to sea. Photo courtesy of Corey Bradshaw.

established. Fewer young pups are surviving their first foraging trips to sea than the number of adults dying each year. Disease, human disturbance and increased numbers of predators have been ruled out as causing this high mortality of the young. The problem seems to be one of food supply. A combination of information gathered on the foraging behaviour, physiology and diet of female seals shows that the fatter a female is when she comes ashore to give birth the better the chance her pup has of surviving its first crucial year. When they do not get enough to eat females carry less fat, have lighter pups, produce less and poorer quality milk, and many fewer young pups will survive. It was thought that competition from commercial fishing might be preventing female seals from getting enough to eat. However, the estimated consumption of fish each year by the Macquarie Island seal population alone is one thousand times greater than the total commercial catch to the south of Australia and New Zealand.

It would seem that changing temperatures and currents associated with changing weather patterns like the El Niño-Southern Oscillation are altering the abundance of phytoplankton and ultimately of the fish and squid on which the elephant seals feed.

Weather can affect food quality

The ways in which weather controls the changes in abundance of other animals can, however, be even more subtle and indirect than any of the foregoing examples. In the case of many herbivorous insects it does so by changing the quality, rather than the quantity, of the insects' food.

Right at the beginning of this book I related how I had been fortunate enough to see the effect of the massive outbreaks of spruce budworm on balsam fir, their preferred food plant, in eastern Canada. And as I later explained, the caterpillars of this small moth have a slightly unusual life cycle. They do not feed when they first emerge from the egg but are dispersed on the wind and then hibernate over winter after moulting to the second stage, still without having fed. When they emerge in the spring they start to feed for the first time, but in the *old* fir needles from the previous year's growth – they start life as senescence-feeders.

Outbreaks of spruce budworm are a feature of large tracts of old, overmature trees, not young, vigorously growing ones. Typically, attacked trees nearly all die after being repeatedly defoliated by immense numbers of caterpillars. The subsequent regeneration of seedlings produces a new even-aged forest in which all trees will become senile at the same time. Modern forest practice attempts to minimise outbreaks by rotational harvesting of trees before they become too old.

But between 1948 and 1958 there was an exceptionally widespread and heavy outbreak, far worse than anything previously experienced. In an attempt to save huge tracts of valuable forest, widespread and repeated aerial spraying with DDT was carried out.

After 1958 the outbreak ceased, including in forests that had not been sprayed, and all seemed well again. Large areas of forest containing millions of trees had been saved, although this meant that a greater proportion of those trees were overmature – and becoming more so by the year. Then, quite unexpectedly, in only 12 years – the shortest interval ever recorded – outbreaks started again.

What is the explanation of all this – and what has it got to do with the weather?

Overmature trees are more susceptible to attack by the budworm because the tissues of their old needles break down more quickly, boosting the level of amino acids in the diet of the young caterpillars which first feed in them. Even in the middle of an outbreak, however, 60 per cent of these still die within days. To lose 60 per cent of the few that find a suitable tree in the first place is fairly drastic mortality, but compared to well over 90 per cent of them dying during times when there is no outbreak, it will quickly produce huge numbers. It is not hard to imagine how a small improvement in their nutrition while feeding on the needles of aging trees could dramatically increase their survival. Yet while this aging of the trees may be a necessary condition, it is clearly not sufficient; old trees last for many years without any sign of being defoliated.

This is where the weather comes into the story. The period 1948 to 1958 was a time when there was a marked departure from the usual pattern of rainfall in eastern Canada. Specifically, there was a series of summer droughts interspersed with much wetter than normal winters. As any gardener knows, if you let the roots of a tree get too dry in summer and then waterlog them in winter, the tree is going to get sick. When this happens its crown dies back as its leaves age and die more quickly, in the process translocating more soluble amino acids from their tissues more quickly than usual.

This double stress of the roots imposed on balsam fir trees that are already over-aged and declining in vigour apparently supplies this extra boost to the diet of young caterpillars feeding in the senescing leaves. It is sufficient, apparently, to raise their survival to a level that can create an outbreak. Saving these trees by killing the caterpillars with DDT meant that they were getting ever more overmature, and thus an improving diet for the young caterpillars, and so more susceptible to attack. At the same time the older they get the more susceptible they become to even quite small perturbations of the weather. So it is not too surprising that outbreaks became more frequent.

A similar – yet different – story has emerged here in Australia for some species of lerp insects feeding on the foliage of eucalypt trees. These species are, like the early-stage caterpillars of the spruce budworm, senescence-feeders. I have already described their life cycle and mode of feeding in Chapter 2. They differ from the budworm in that they feed on old, dying tissues for the whole of their immature stage, not just the first part of it. And they even hurry this process along so that the bit of leaf on which they feed dies before the rest of the leaf, so releasing more good food faster than it would otherwise have done. Even so, they remain pretty rare for most of the time – often so much so that they are almost impossible to find. Clearly, then, these adaptations to improve their chances of surviving and multiplying have not been all that successful because for most of the time they just manage to persist in their habitat. Yet now and again they become so abundant that nearly every leaf, on every tree, over many square kilometres of countryside, is covered with hundreds of them. Then most of the leaves die and are shed prematurely; in extreme outbreaks to the extent that all the trees are stripped bare of foliage. It eventuates that, as was the case with the spruce budworm, these outbreaks occur at times when there have been a series of much drier summers and much wetter winters. Apparently this pattern of weather puts the trees under sufficient stress that their crowns will start to die back. And this means the process of aging of the leaves and the concomitant breakdown and export of nutrients from them is speeded up. Adding this to the capacity of the young insects to accelerate the rate of breakdown of the tissues that they feed from would seem to be sufficient to tip the balance in favour of most of them surviving, whereas most of the time most of them die soon after they start to feed. In a typical life cycle of three generations in 12 months this quickly results in explosive growth of their numbers. If this pattern of weather persists over just a few years, an outbreak of untold millions of insects ensues. But once the weather again returns to more normal patterns, few young lerp insects survive their first couple of days of life, and the trees quickly regenerate vigorous and lush new growth wherein it is once more exceedingly difficult to find a lerp insect.

So we see the same pattern emerge for a wide variety of both herbivores and predators. Changes in the weather generate changes in the availability of protein food that is sufficient to sustain breeding and growth of the young. The numbers of the animals rise and fall in response to this. One minute they are rare, the next in huge numbers, and finally again rare – as most die as the good food disappears.

In all of these stories great increases in abundance – outbreaks, plagues, epidemics, call them what you will – provide the unusual event which confirms the usual condition. In the exception one discovers the rule. Animals are adapted to make maximum use of whatever resources are available to them

– in particular, food that will support the production and growth of new young. They are at all times pressing hard against the limits set by the amount of this food that they can gain access to. However, the variability of the weather is such that shortage is the norm, abundance an unpredictable rarity. Hence animals that persist in their normally inhospitable world are those that have become sufficiently adept at finding and using food in that world. Even so, it is a constant struggle, and only a few manage to succeed. Consequently, their inherent response to the exceptional event of a greatly increased supply of food is to use as much of it as they can as quickly as possible – and breed explosively. And it is this capacity to respond immediately to any alleviation of their lot which illustrates that during 'normal' times the animals are few in number because they are limited by the amount of good food they can obtain, not by any other factor.

Whether you are an insect, a bird, a reptile, a mammal – or a bacterium or a fungus, if you have evolved and are programmed to persist in a world which is inadequate for most of the time, then if that world occasionally – and unpredictably – becomes very benign, you are going to respond by increasing your numbers very quickly; and will keep on doing so as long as the good times last. But when, sooner or later, things return to 'normal', your numbers will quickly return to those few that can eke out an existence in the usual harsh world.

'What is the point?' you may ask. 'They are only going to die in the end.' But that is the wrong question to ask – and the wrong conclusion to draw. Nature does not reason. Nor does it have any 'purpose'. Each individual organism will maximise the number of genes that it can pass on to the next generation. To do so each is programmed (by its genes!) to use each and every opportunity to increase the number of its offspring. And the only way to do this is to maximise the amount of nutrients that can be extracted from the environment and converted into more breeding individuals.

It is necessary to understand that animals do not live in a benevolent world wherein all they need to live out full and happy lives is readily to hand. For most of the time the reverse is true. They live in a world which is mostly harsh and unforgiving, and where food of sufficient quality is nearly always in short supply. It is failure to realise this which leads people to conclude that it is only because their predators kill so many of them first that animals do not become so frequent that they destroy all their resources.

To persist – to not become extinct – in a world where times are nearly always tough, you must have evolved traits which enable you to be very efficient at finding what resources are available to you and in using them as efficiently as possible. All the numerous anatomical, behavioural and physiological specialisations I have related in this book help do this.

There is one other species of animal which illustrates all this very well – our own. Until about 10 000 years ago humans, while they had been successful enough to spread to most parts of the world, lived as small bands of people constantly moving across the land. They were hunter-gatherers, surviving on what little meat they could kill or scavenge (most often the latter), and harvesting fruits, seeds and roots. They were entirely at the mercy of uncontrollable events in their environment – the movement of herds of animals and the seasonal growth of plants. And these in turn were entirely determined by the weather. Ten thousand years ago, however, the weather was becoming warmer and wetter as the great icesheets receded at the end of the last ice age. Coincident with this, people started to become farmers; they learnt to control the production of their food. They began to domesticate wild plants and animals. This happened gradually and independently at different times and places (for example, beans and cattle in India, maize and llamas in South America, rice and pigs in China). But it was an irreversible and inexorable process once started, enabling the productivity of the 10 or so square kilometres needed to support one hunter-gatherer to be increased by as much as 50 times. And it was not control of the production of just any old food, but of the high quality protein foods – meat and seeds – that are vital for successful breeding and nurturing of the young.

The result is all too well known: the greatest – and longest lasting – outbreak of all time, far exceeding anything previously experienced by any other organism. But such success has its costs, on two fronts.

First, the growing of large amounts of genetically less-diverse plants, in greater and greater concentrations, and learning how to improve and extend their growth by the addition of fertilisers and water, provided a huge increase in the amount and quality of food available for other organisms which had evolved to feed on those plants in their pre-domestication days. Imagine, if you will, the response of grain weevils evolved to hunt out and lay eggs in the few seeds of scattered grass plants, now confronted with field upon field of them, and all bearing many more, bigger and more nutritious seeds than their ancestors. Or silos full of those same seeds. The same is true for domestic animals. A carnivore like a lion or dingo, adapted to hunt for small numbers of lean and swift prey scattered widely throughout its environment, will quickly learn to switch to the 'easy meat' of confined herds of the much fatter and less mobile specimens being bred and fed by humans. Similarly for the highly destructive carpet beetles, evolved to find and eat the skin and hair on occasional animal carcasses not already eaten by other scavengers. Presented with huge concentrations of wool in carpets and clothing they quickly multiply and consume large amounts of this new-found resource. The same thing has happened with all predators, parasites and diseases of domesticated plants and animals. Predictably, such organisms have flourished, producing large and persisting outbreaks in response to the huge increase in their food. They have become the pests we must constantly be combating to save our farmed resources for ourselves.

Second, the huge increase and concentration of populations of ever-better-fed humans meant the proliferation of our own predators, parasites and diseases. Epidemics – outbreaks – become inevitable, only subsiding when natural selection kills most of the susceptible people, leaving only resistant individuals to breed (tragic illustrations of this have been the devastation wrought by smallpox, syphilis and measles introduced by Europeans to indigenous populations not previously exposed to these diseases). More recently we have learned to kill such organisms. They, however, inevitably, evolve resistance to our killing agents (think of the problem of 'golden staph' in our hospitals), so we must be constantly devising new ways to combat them.

And there is a final point about this outbreak of humans. Whereas, like all other animals, the outbreak clearly has been generated by access to increased amounts of good food, this increase was not generated by the weather in the way it is for outbreaks of other animals. We have overcome the tyranny of the weather. Obviously this is not entirely true. Exceptional weather like floods and droughts can make serious inroads into our supplies of food. But these are relatively minor blips. Our overall access to food is now not a function of the longer-term fluctuations of the weather.

Afterword

I specifically discussed the inefficiency of predators in Chapter 6, but the rest of the book illustrates that herbivores are just as inefficient at 'controlling' their 'prey'. Predators are inefficient because their prey have evolved ways to mostly stop being caught – to become inaccessible. If this had not happened then the end point would have been extinction of both prey and predator. Similarly plants have evolved, not so much to be inaccessible, but to be nutritionally inadequate for their predators – the herbivores – thus reducing them to similar levels of inefficiency, but once more avoiding mutual extinction.

We sometimes see speeded-up versions of this co-evolution in action as a result of human interference with the natural world. An especially good example happened here in Australia. When the myxomatosis virus was introduced to attack the devastating hoards of feral rabbits it swept through their populations, via its mosquito vector, reducing their numbers by over 90 per cent. The two organisms had never before encountered each other in nature, so rabbits were a new and untapped food resource for the virus; the virus a new predator against which the rabbit had no defence. But in relatively few years a double phenomenon emerged. The few rabbits that survived did so because they had some resistance to the virus. They quickly bred to considerable numbers, only to be once more decimated by the disease – but leaving a slightly larger and even more resistant residual population. And this happened many times, on each occasion with fewer and fewer rabbits dying. Simultaneously, however, the virus attacking each generation consisted of a less virulent strain than before, one less likely to kill a rabbit before a mosquito could transmit it to a new host. The end point is that rabbits are little affected by myxomatosis any more, and we have had to introduce another virus – calicivirus – to start the process all over again.

Throughout this narrative I have stressed that it is the food that is available to individual animals in a population which is the crucial element deciding how many of them will live. Much of the potential food in a habitat is not used because it is quite inaccessible. Whether it is because it can run too fast for you to catch, or it is too dilute in your diet of plant sap, is of little consequence. In either case you will die for lack of food in the midst of apparent plenty. And this has come about because of the constant co-evolution of eater and eaten.

Another thing that prevents animals gaining access to all the potential food in their environment is evolution's 'one way arrow': once launched down a particular evolutionary path there is no turning back. The senescence-feeding lerp insects I have discussed before provide a good example of how this works. They will feed only on mature gum leaves even when, within easy reach on the same twig, there are young expanding leaves containing much higher concentrations of amino acids than the mature leaves. Even when I confined them experimentally to new leaves they would not feed or lay their eggs on them, and newly hatched nymphs placed on new leaves wander until they die rather than feed. These insects are descended from ancestors that happened (for reasons we can never know) to feed on old leaves and survive. Now they are locked into that way of life by their inherited physiology and behaviour. They will respond only to specific cues from mature leaves, so new leaves might just as well be on another planet; they are not part of their world, and the food in them, no matter how good, is not available to them. In the same way cabbage white butterfly caterpillars will die without feeding if confined to a plant that is not a brassica, even though it might contain more food than the brassica.

So, to say a population is limited by its food does not infer that it eats all the food that is there. There is often lots left, but they cannot get at it. That there is this reservoir of unused food in many habitats is dramatically demonstrated from time to time when human activity results in an animal getting into a part of the world where it had never been before. There are many examples, especially in this part of the world – rabbits, wasps, millipedes, possums, pigs, goats, deer; to name but a few. Typically the numbers of such new introductions explode at the point of introduction and the population rapidly spreads. Conventional wisdom says this is because they no longer have their natural enemies to control their numbers. But I hope you have learnt enough from this book to realise that this is unlikely to be the answer. Far more probable is that the population has found a source of unexploited food that was not accessible to the native animals. A phenomenon repeatedly observed with such new introductions indicates this might be the case. It is the 'doughnut effect'. As a new population becomes established it increases to very high numbers before starting to spread. Then, as it spreads, its numbers at the point of origin drop markedly while at the 'wave front' moving out in all directions from the source, they remain large. They exhaust and overrun the hitherto unused supply of food as they spread. Another theory is that the invader actively out-competes and excludes whatever native animal was using the resource.

Mostly such conclusions are based on anecdote or untested assumptions. And in some cases they may well be true. However, the few studies that have looked carefully for this result have found it not to be the case. One such was done here in Adelaide. The recently arrived European wasp and the native Australian paper wasp both hunt insect prey which they feed to their growing young in their paper nests. It was feared the much more abundant and aggressive newcomer would quickly oust the native wasp by aggressively beating it for its food. However, it has not done so and monitoring the day-to-day activities of the two species demonstrated why. They hunt in different places within the same habitat, at different times and over a different range of ambient temperature. The differences in time and temperature relate to differences in the climate of their native habitats. But the difference in their hunting relates to their exploiting different food. The native wasps hunt exclusively in the foliage of plants for caterpillars. The invaders, on the other hand, are voracious omnivores (and scavengers – as anyone trying to have a garden barbecue with these wasps around will attest). They prey on a wide range of insects in many locations, but mostly catch adult flies. Caterpillars, however, make up less than 8 per cent of their diet, and it is probable that the ones they do take belong to different species from those the paper wasps catch.

A similar story, but with a much longer history, emerged when the European white butterfly was introduced into the eastern United States in the 19th century. It spread rapidly and reports from naturalists and casual observers indicated it was ousting a closely related native American white butterfly which was becoming increasingly rare and was thought would soon be exterminated all together. Yet 100 years later both species were abundant and living in the same locations. Close study of their respective life cycles revealed, however, that each species lives in a different world; their habitats are different. They eat different species of plants, and even when adult butterflies of both are flying and mating in the same field, they completely ignore each other. It seems the great success of the invader and the local extinctions of the native came about because major changes in land use with early European farming proliferated the preferred plants of the European butterfly while destroying much of those required by the native one.

Figure A.1 Polistes wasp: Fears that these native Australian paper wasps would be supplanted when the introduced European wasp ate most of their prey proved unfounded. The two species have less than 8 per cent of their prey in common. Photo courtesy of Kym Perry.

Where then, at the end of this book, does this leave us? I would hope understanding why the world stays green; that it does so because all animals are pretty much struggling to survive. They live in a world where they are usually hard pressed to find enough food to persist, in spite of the myriad ways they have evolved to overcome this shortage. This is not because there isn't enough food in the world, but because much of it is unavailable, is spread too thinly, or is too hard to catch. From a consideration of all this it should not be difficult to appreciate that the idea of an efficient and 'balanced' nature is just a myth. Animals do not make optimal use of their habitat, but rather have evolved ways of maximising their access to what resources are present. Depending on the quirks of natural selection, this may or may not be the 'best' outcome as seen from our human standpoint. By and large nature is just coping – it gets by with what works at the time.

Further reading

More detailed information about many of the stories related here (and of a good many others) can be found in my 1993 book, *The Inadequate Environment*, and its 1150 references to the ecological literature. Since 1993 scientists have published many new stories, and I have incorporated a lot of them into this book. But they can be found only in the scientific journals. The reader who wants to delve more deeply will have to go directly to those journals. To aid those who wish to do so, I have listed a selection of articles, including some not discussed in the text, which will enable them to follow up more recent work in the topic of each chapter.

Chapter 1
Den Boer, P.J. (1999). Natural selection, or the non-survival of the non-fit. *Acta Biotheoretica* **47**, pp. 83–97.

White, T.C.R. (1993). *The inadequate environment: nitrogen and the abundance of animals.* (Springer-Verlag: Berlin, Heidelberg, New York.)

White, T.C.R. (2001). Opposing paradigms: regulation or limitation of populations? *Oikos* **93**, pp. 148–52.

Chapter 2
Allen, L.R. & Hume, I.D. (2001). The maintenance nitrogen requirement of the Zebra Finch *Taeniopygia guttata*. *Physiol Biochem Zool* **74**, pp. 366–75.

Thompson, V. (2004). Associative nitrogen fixation, C4 photosynthesis, and the evolution of spittlebugs (Hemiptera: Cercopidae) as a major pest of neotropical sugar cane and forage grasses. *Bull Ent Res* **94**, pp. 189–200.

White, T.C.R. (2002). Outbreaks of house mice in Australia: limitation by a key resource. *Aust J Agric Res* **53**, pp. 505–9.

Chapter 3
Dellomo, G., Alleva, E. & Carere, C. (1998). Parental recycling of nestling faeces in the common swift. *Anim Behav* **56**, pp. 631–8.

Kenagy, G.J., Veloso, C. & Bozinovic, F. (1999). Daily rhythm of food intake and faeces reingestion in the Degu, an herbivorous Chilean rodent: optimising digestion through coprophagy. *Phys & Biochem Zool* **72**, pp. 78–86.

Nalepa, C.A., Bignel, D.E. & Bandi, C. (2001). Detritivory, coprophagy, and the evolution of digestive mutualisms in Dictyoptera. *Insect Soc* **48**, pp. 194–201.

Trewick, S.A. (1999). Kakapo: the paradoxical parrot. *Nature Australia* **26**, pp. 54–63.

Chapter 4

Barros-Bellanda, H.C.H. & Zucoloto, F.S. (2001). Influence of chorion ingestion on the performance of *Ascia monuste* and its association with cannibalism. *Ecol Entomol* **26**, pp. 557–61.

Brune, A. & Kuhl, M. (1996). pH profiles of the extremely alkaline hindguts of soil-feeding termites (Isoptera; Termitidae) determined with microelectrodes. *J Insect Physiol* **42**, pp. 1121–7.

Caldwell, J.P. (1997). Pair bonding in spotted poison frogs. *Nature* **385**, pp. 211.

Dudley, J.P. (1998). Reports of carnivory by the common hippo *Hippopotamus amphibius*. *Sth Afr J Wildl Res* **28**, pp. 58–9.

Gonzales-Pastor, J.E., Hobbs, E.C. & Lusick, R. (2003). Cannibalism by sporulating bacteria. *Science* **301**, pp. 510–13.

Hollingham, R. (2004). Natural born cannibals. *New Scientist* **183**, pp. 30–3.

Hurd, L.E., Eisenberg, R.M., Fagan, W.F., Tilmon, K.J., Snyder, W.E., Vandersall, K.S., Datz, S.G. & Welch, J.D. (1994). Cannibalism reverses male-biased sex ratio in adult mantids: female strategy against food limitation? *Oikos* **69**, pp. 193–8.

Mead, S., Stumpf, M.P.H., Whitfield, J., Beck, J.A., Poulter, M., Campbell, T., Uphill, J.B., Goldstein, D., Alders, M., Fisher, E.M.C. & Collinge, J. (2003). Balancing selection at the prion protein gene consistent with prehistoric kuru-like epidemics. *Science* **300**, pp. 640–3.

Milne, M. & Walter, G.H. (1997). The significance of prey in the diet of the phytophagous thrips, *Frankliniella schultzei*. *Ecol Entomol* **22**, pp. 74–81.

Mira, A. (2000). Exuviae eating: a nitrogen meal? *J Insect Physiol* **46**, pp. 605–10.

Moir, R.J. (1994). The 'carnivorous' herbivores. In: *The digestive system in mammals: food, form and function*. (Eds D.J. Chivers & P. Langer) pp. 87–102. (Cambridge University Press: Cambridge.)

Preen, A. (1995). Diet of dugongs: are they omnivores? *J Mammal* **76**, pp. 163–71.

Zamora, R. & Gomez, J.M. (1996). Carnivorous plant-slug interaction: a trip from herbivory to kleptoparasitism. *J Anim Ecol* **65**, pp. 154–60.

Zamora, R., Gomez, J.M. & Hodar, J.A. (1997). Responses of a carnivorous plant to prey and inorganic nutrients in a Mediterranean environment. *Oecologia* **111**, pp. 443–51.

Chapter 5

Ochiai, K. & Susaki, K. (2002). Effect of territoriality on population density in the Japanese Serow (*Capricornis crispus*). *J Mammal* **83**, pp. 964–72.

Southwick, C.H. (1955). The population dynamics of confined house mice supplied with unlimited food. *Ecology* **36**, pp. 212–25.

Strecker, R.L. & Emlen, J.T. (1953). Regulatory mechanisms in house-mouse populations: the effect of limited food supply on a confined population. *Ecology* **34**, pp. 375–85.

Chapter 6

Beckman, N. & Hurd, L.E. (2003). Pollen feeding and fitness in praying mantids: the vegetarian side of a tritrophic predator. *Environ Entomol* **32**, pp. 881–5.

Blackwell, G.L., Potter, M.A., McLennan, J.A. & Minot, E.O. (2003). The role of predators in ship rat and house mouse population eruptions: drivers or passengers? *Oikos* **100**, pp. 601–13.

Crivelli, A.J. (1994). Why do white pelican chicks die suddenly on Arel Island, Banc d' Arguin in Mauritania? *Rev d' Ecol la Terre et la Vie* **49**, pp. 321–30.

King, C.M., White, P.C.L., Purdey, D.C. & Lawrence, B. (2003). Matching productivity to resource availability in a small predator, the stoat (*Mustela erminea*). *Can J Zool* **81**, pp. 662–9.

Chapter 7

Bradshaw, C.J.A. & Hindell, M.A. (2003). Fat explorers of the deep. *Nature Australia* **27**, pp. 34–43.

Jaksic, F.M., Silva, S.I., Meserve, P.L. & Gutierrez, J.R. (1997). A long-term study of vertebrate predator response to an El Niño (ENSO) disturbance in western South America. *Oikos* **78**, pp. 341–54.

Madsen, T. & Shine, R. (1999). Rainfall and rats: climatically-driven dynamics of a tropical rodent popualtion. *Aust J Ecol* **24**, pp. 80–9.

McMahon CR and Burton HR (2005) Climate change and seal survival: evidence for environmentally mediated changes in elephant seal, *Microunga leonina* pup survival. *Proc R Soc B* **272**, pp. 923–8.

Newsome, A.E., Catling, P.C., Cooke, B.D. & Smyth, R. (2001). Two ecological universes separated by the dingo barrier fence in semi-arid Australia: interactions between landscapes, herbivory and carnivory, with and without dingoes. *Rangl J* **23**, pp. 71–88.

Packer, C., Hilborn, R., Mosser, A., Kissul, B., Borner, M., Hopcraft, G., Wilmshurst, J., Mdura, S. & Sinclair, A.R.E. (2005). Ecological change, group territoriality, and population dynamics of Serengeti lions. *Science* **307**, pp. 390–3.

Sykes, B. (2001). *The seven daughters of Eve*. (Bantam Press: London.)

White, T.C.R. (2004). Limitation of populations by weather-driven changes in food: a challenge to density-dependent regulation. *Oikos* **105**, pp. 664–6.

Zann, R.A., Morton, S.R., Jones, K.R. & Burley, N.T. (1995). The timing of breeding by Zebra Finches in relation to rainfall in Central Australia. *Emu* **95**, pp. 208–22.

Index

1080 (poison) immunity   6
aerial spraying *see* insecticides
altruism   70
amino acids   7, 10, 16, 17, 18, 19, 20, 22, 25, 26, 29, 34, 37, 45, 55, 67
ant-lion   81–82
ants, 81; white 36–37, 49, 62; wood 59–61
aphids, 5, 22, 24, 26, 30, 45, 67, 79–80; Phylloxera 30; rose 16; spruce 28; sycamore 29
arthropods   55
baboons   71
bacteria, 33, 44, 45; caecal 41, 42; cannibalistic 59; endosymbiotic 34, 55–56; nitrogen fixing 25
bamboo   23
barley   20
bats   23–24, 48
bears   70–71
beech   22, 30, 91, 101
beetles, ground   87
boom and bust fluctuations   89
bream, buffalo   38
budworm, spruce   1, 2, 31, 32, 104–105, 106
bumblebees   85
butterflies, cabbage white 6, 15, 23, 112; checkerspot 30–31; wanderer 6; white 113
C:N ratio   10
caeca   34, 39, 41, 42
calicivirus   111
cannibalism, 53, 54, 55, 56–62, 76; embryo 58; neonate 11, 51–54, 55, 57, 61, 103, 107; sibling 58, 59; strategies 58–59
capercailzie   51
carbohydrate   7, 11, 24, 26
carbon   9, 10, 11
carnivores   11, 50, 51–54, 62, 79, 90, 98
carnivorous plants   10, 54
carnivory, insects 61–62; opportunistic 47
cast skin   55–56, 57
caterpillars   1, 2, 19, 22, 25, 29–32, 55, 57, 61, 80, 104–106, 112
cats (feral domestic)   82, 83, 97
cattle   50
cellulase   33
cellulose   11, 16, 23, 33, 34, 36, 38, 41, 43, 44, 45
chimpanzees   49
chyme   42
cichlids   53
climate   93–109
cockatoos   17–18, 20
cockroaches   37–38, 55–56

competition   8
condors, Andean   65–66, 100
cooperative breeders   72–73
coprophagy   33–42, 43, 44
corellas   52
crabs   54, 59
creaming off   23, 24, 40
crickets   54–55
crop milk   53
crops, human   3
cuckoo, Galapagos   95
decomposers   3, 12, 33, 44, 45
deer   50, 71, 83
defence mechanisms, plants   5–6
deterrent chemicals   6, 16
detritus-feeders   43–45
dingo fence   97
dingoes   96–97
diseases   104, 108, 109, 111
dispersal, aerial 5, 32, 104; insect 5, 32, 104
doomed surplus   67, 68, 70, 74
doughnut effect   112
drought   66, 83, 95, 98, 105
ducks   51, 52, 64
dugong   48, 50
dung-eaters   33–42
eagles   98
egg-shell eating   55, 56, 62
eggs, 51, 53, 55; trophic 54, 57
El Niño   64, 66, 89, 94, 95, 100, 104
elephant seals, southern   103–104
embryo cannibalism   58
energy   9, 10, 11, 17
ENSO   95, 103, 104
epidemics *see* outbreaks
Eucalyptus   5, 15-16, 26, 53, 106
faeces, caecal 39–42; faeces eating 33–45; gorilla 49; lerp 26–27; panda 50; pellets 39, 40, 41, 42, 44; volume of 7, 24
fast-track feeders   23–26
favoured few   70
fecundity   102
feeding behaviour   1–5, 15–32
finches, 93–95; gold 18, 42; large cactus 64, zebra 18, 19, 99
fir, balsam 1, 2, 31, 32, 104, 105; silver 22
fish   33–34, 38, 53
flies, 20, 21, 28, 87; caddis 21; saw 28; tsetse 87
floaters *see* doomed surplus
flush-feeders   15–17, 25, 27, 28, 30, 32
food, faeces as *see* coprophagy; fast passage of 23, 24; identifying in gut/faecal samples 49,

50, 52, 61; poisonous 5–6; pollen as 11, 24, 35, 85, 86; poor quality 7, 104–109; relative vs. absolute shortage of 10, 11, 80, 86, 87; taste 5–6
food availability   70, 77, 88–91, 95, 111
food supply, weather-driven   98, 99, 104–109
fore-gut fermenters   34, 35, 36
foxes, Arctic 83; red 64, 74, 83, 90, 97
frogs   54
fungi   37, 44, 45
fussy eaters   15–32
galahs   18
galls   21–22, 30, 48, 67
geese   21, 24, 52
genes   8, 59, 65, 70, 73, 75, 107
giraffes   21
glider, sugar   48
goats (feral)   71
gorillas   19, 49
grasshoppers   54–55, 62
grasslands   1, 52, 68
grazer strategy   58, 59
grazers   20, 34–35, 44, 48, 52, 63
grouse   21, 40–42, 51, 99
gut micro-organisms *see* micro-organisms
hares   21, 39, 82
hatching, asynchronous   55
hatching, helpers *see* cooperative breeders
heather   21
herbivore–micro-organism associations   33–45
hind-gut fermenters   34, 36, 38–42
hippopotamus   49–50
hoatzin   35
honeydew   24, 26
honeyeaters   52
horse   38–39
host-specific species   15, 16
humans, 61, 71, 107–109; refuse 70–71, 82
hummingbirds   52
hyenas   98
iguanids   40, 53, 95
immatures *see* neonates
insecticides   1, 105
insectivorous plants   10, 54
introduced predators   90–91
invertebrates, bungling   84–87
kakapo   35–36
kangaroos   34, 96–97, 100
kiwi   91
koalas   15–16, 40
kookaburra   65
leafhoppers   28–29
leafminers   29–30
legumes   15, 25, 35, 40
lemurs   49
lerp insects   26–28, 106, 112

lifeboat strategy   58–59
lignin   11, 36
limpets   21, 63–64
lions   80–81, 98
locusts   1, 3, 29, 62, 98
lorikeets   24
lynx   71, 82
magpies   66
malleefowl   90
malnutrition *see* starvation
mantids   57, 58, 85–86; Chinese 86; praying 57, 58, 85–86
marine shipworms   44–45
marmots   73–74
mast years   35–36, 101
mice, 20, 47, 95; feral house 20, 48, 75–77, 91, 101; sandy inland 48; spinifex hopping 48
micro-organisms, 3, 12, 44, 45, 62; caecal 34; gut 33, 37, 38, 42, 43; nitrogen-fixing 25
microtines *see* voles
milk   40, 51, 57, 58, 81, 101, 102, 104
milk, crop   53
milk-ripe seeds   17, 18, 19, 42
mistletoe bird   52
mites   5, 56
mockingbird, Galapagos   95
molluscs   54
monkeys   49
moths, 23, 29, 47, 61, 104; codling 19–20, 59, 67
mycetome   45
myxomatosis virus   111
natural enemies *see* predators
natural selection   13
nectar   52
neonates, 18, 40, 42, 52, 53, 56, 62, 65, 67, 72, 80, 83, 89, 95, 99, 106; cannibalism 11, 51–54, 55, 57, 61, 103, 107; mortality 75, 95, 101; starvation 6, 11, 34, 50, 57, 59, 62, 81, 82, 83, 86, 85, 89, 91, 98; surplus 12–13
nesting sites, limiting resource   64–65
nestlings *see* neonates
nitrogen, 7, 9–13, 19; fixing 25, 34, 36, 45; metabolic 34, 36, 42; recycled 23, 36–37, 41, 55, 56; soluble 21, 40
non-survival of the non-fit   13
nutrients   4, 6–7, 107
offspring, surplus   12–13
omnivores   33, 41, 48, 49, 55, 113
outbreaks   3, 25, 32, 91, 95, 98, 104–106, 108–109
panda, giant   23, 50
pap, koala   40
parasites   22, 80, 108, 109
parrots   18, 35–36, 52–53
pelagic larvae   53
pelicans   88–90, 100

pests 3, 25, 109
photosynthesis 9
phytoplankton 104
pigeons 53
pines 1, 15, 18, 25, 35, 64
plagues *see* outbreaks
plant life 3
plants, defence mechanisms 5, 16
pollen 11, 24, 35, 85, 86
poplars 30, 67
population cycles 29, 69, 71, 72, 74, 75–77, 88–90, 112
population, limited by food 112, 114
possums, brushtail 48; ringtail 39, 40
predators, 4, 8–9, 48–50, 65, 79, 83, 86, 90–91, 111; opportunistic 54–56
primates 74
protein 10, 11, 16, 17, 18, 19, 23, 24, 34, 37, 38, 40, 51, 53, 55–56, 61, 62, 79, 99, 106
protozoans 34, 37, 40
psyllids 26–28, 45
ptarmigan 51
python, water 102
rabbits 39, 64, 74, 82, 83, 90, 91, 97–98, 111
rain, lack of 66, 96, 98, 105
raptors 59, 65
rats, black 76, 91, 95; dusky 101–102, 103; kangaroo 65
reproductive capacity 9, 13, 50, 76, 88, 95, 100, 102, 103
reproductive strategies 100–104
reptiles 51, 53
rice 20, 29
rodents 39–40, 48, 73, 91, 95, 101–102
rubbish dumps 70–71, 82
Rubisco 62
Rumen, 34, 35, 43, 45; external 43
salamanders 58
sawflies 28
scale insects 22, 28
scorpions 59, 60
seagrass 3, 21, 34, 48, 53
seed-eaters 17–20, 52, 53, 94, 99
seeds, 17, 18, 19, 21, 36, 94, 95, 96, 101, 108; cereals 20, 29; immature 17, 18, 20, 35, 59, 67, 99; mast 91, 101; milk-ripe 17–19, 42
selective feeding 16
senescence feeders 25–32, 104, 106, 112
serow 68–69
sharks 58
sheep 21, 50
shipworm 45
siblicide 67
sibling, cannibalism 58, 59

slater *see* woodlice
slugs 54
snails 44, 54, 57, 63
snakes 102
social dominance hierarchies 71–77
soluble nitrogen *see* nitrogen, amino acids
specialist feeders 6, 16, 35, 43, 49, 52
spiders, 57, 68; jumping 85; katipo 79; orbweb 79, 84–85
spittle bugs 24–25
spores, bacterial 59; fungal 37
sporulation 59
squirrels 19, 48
starvation 6, 11, 34, 50, 57, 59, 62, 81, 82, 83, 86, 85, 89, 91, 98
stoats 71, 91, 100–101
struggle for existence 12, 13
surplus young *see* doomed surplus
survival of the fittest 13
swift, European 42
takahe 51–52
tannins *see* deterrents
termites 36–37, 49, 62
territorial behaviour, 49, 50, 63–77; insect 67–68
thrips 56
ticks 86
titmice 64
tortoise, Aldabra 21
toxins 6
trophic eggs 54, 57
tsetse fly 87
turtles, green 21, 53
universal nitrogen hunger 12
upwelling, ocean 89, 100
vectors, mosquito 111
vegetarians, meat-eating 48–56; obligate 49
voles 39, 49, 83, 101
wallabies 47–48
walnut 18, 35
warbler, Seychelles 72–73
warfare, human 61; wood ant 60–61
wasps, 68, 80; European 113; paper 113; parasitic 80; spider-hunting 79–80
weather, affects of 93–100, 105, 106
weather-driven food supply 98, 99, 104–109
weeds 18, 20
weevils 30
wildebeest 98
wolverines 71
wolves 71, 83
woodlice 43–44
yolk sac 51
young, very *see* neonates